Power Systems

For further volumes:
www.springer.com/series/4622

Surajit Chattopadhyay • Madhuchhanda Mitra
Samarjit Sengupta

Electric Power Quality

Springer

Surajit Chattopadhyay
Electrical Engineering Department
Hooghly Engineering and Technology College
West Bengal University of Technology
Hooghly, West Bengal
India
surajitchattopadhyay@gmail.com

Madhuchhanda Mitra
Department of Applied Physics
University of Calcutta
92 APC Road
Kolkata 700009, West Bengal
India
madhuchhanda94@rediffmail.com

Samarjit Sengupta
Department of Applied Physics
University of Calcutta
92 APC Road
Kolkata 700009, West Bengal
India
samarsgp@rediffmail.com

ISBN 978-94-007-0634-7 e-ISBN 978-94-007-0635-4
DOI 10.1007/978-94-007-0635-4
Springer Dordrecht Heidelberg London New York

Library of Congress Control Number: 2011921328

Printed on acid-free paper

Springer is part of Springer Science+Business Media (www.springer.com)

Foreword

Electrical Power has become the life line of our civilization. It is considered as an indicator of the stage of development of a country. The quantitative and qualitative development of the sources of electricity is the most important requirement for the power utility. Various technologies have been developed in case of conventional power generation e.g. thermal, hydel or nuclear. Again some non-conventional energy sources like wind power, solar power or mini-micro hydel power are also contributing to the total power bank. Presently the electricity grid is receiving power from multiple sources, both conventional and non-conventional. This hybrid system requires tight quality control particularly using improved measuring techniques of power quality parameters for this power mix. The energy engineers and technologists are striving hard to find out ways and means to solve the problems related to power systems, due to the mixing of power from various sources. Researchers are carrying out their studies on different aspects of the problem utilizing modern electronic devices, smart sensors and state-of-the-art control protocols.

The present book written by my students, is the result of their prolonged research work in the area of power quality issues. This is a timely publication and will be much appreciated by both undergraduate and postgraduate students. It will also serve as a reference book for the researchers carrying out researches in the relevant areas. It is felt that the content of the book is well organized and innovative.

I must heartily congratulate the authors for the publication of the book. It is hoped that this book will satisfy the requirements of those for whom it has been written.

08 December 2010 — Prof. (Dr.) Dilip Kumar Basu
Kolkata

Preface

Day-by-day electric power systems are becoming more and more complex. The dependence of power system on distributed energy sources, including renewable and non-conventional, has made the control of the system sufficiently intricate. With the use of modern power electronic devices, now-a-days, the complexities in system contrology are made more efficient, user-friendly and reliable also. But the usage of these devices has pushed a power system in serious quality problem. Since the use of sophisticated electronic gadgets has increased in every sphere of life, for their good longevity, requirement of quality power has become a predominant criterion to the consumers in the present deregulated competitive power market. Therefore, electric power quality has become the concern of utilities, end users as well as manufacturers. This book is intended for graduate, postgraduate and researchers as well as for professionals in the related fields.

This book has evolved from the researches carried out by the authors and the contents of the courses given by the authors at University of Calcutta, Department of Applied Physics, India in the Bachelor and Master's courses in Electrical Engineering. A large number of references are given in the book most of which are journal and conference papers and national and international standards.

The contents of the book focuses, on one hand, on different power quality issues, their sources and effects and different related standards, and on the other hand, measurement techniques for different power quality parameters. Advantages and limitations of different methods are discussed along simulated and laboratory experiment results. At the end, a chapter has been added which deals a concept of generation of harmonics in a power system and its components.

The key features of the book can be highlighted as follows:

- This book has approached the subject matter in a lucid language. Measurement techniques have their analytical background supplemented by simulated and experimental results.
- This book has mainly handled with measurement techniques of power quality parameters, which is absent in many other similar books.
- In general, the book has dealt with different power quality issues which are required for students, researchers and practicing engineers.

- The content level of the book is designed in such a way that the concepts of different power quality issues in modern power system are built up first, followed by some existing and new measurement methods. This content should attract the students, researchers and practicing engineers.
- The predominant features of the book are
  - Lucid but concise description of the subject (which may be available in other books).
  - Detailed new measurement techniques (which are not available in other books).

The authors wish to thank members of the Springer publisher of our book.

They owe a particular debt of gratitude to the teachers of Department of Applied Physics for their constant support in preparing the manuscript. At last, but not the least, the authors are indebted to their better-halfs and children, without whose constant endurance it would not have been possible for this book to see the light.

Contents

List Principal Symbols and Acronyms

V_R = Amplitude of voltage in phase R
V_Y = Amplitude of voltage in phase Y
V_B = Amplitude of voltage in phase B
V_{R0} = Zero sequence voltage in phase R
V_{Y0} = Zero sequence voltage in phase Y
V_{B0} = Zero sequence voltage in phase B
V_{R1} = Positive sequence voltage in phase R
V_{Y1} = Positive sequence voltage in phase Y
V_{B1} = Positive sequence voltage in phase B
V_{R2} = Negative sequence voltage in phase R
V_{Y2} = Negative sequence voltage in phase Y
V_{B2} = Negative sequence voltage in phase B
$v(t)$ = Voltage
$v_R(t)$ = Voltage of phase R
$v_Y(t)$ = Voltage of phase Y
$v_B(t)$ = Voltage of phase B
$v_R^N(t)$ = Normalized voltage of phase R
$v_Y^N(t)$ = Normalized voltage of phase Y
$v_B^N(t)$ = Normalized voltage of phase B
v_{R1} = Fundamental component of R phase voltage
V_{Rm} = Harmonic component of R phase voltage
$V_{\alpha 1}$ = Amplitude of fundamental components of voltage of α axis
$V_{\beta 1}$ = Amplitude of fundamental components of voltage of β axis
$V_{\alpha m}$ = Amplitude of mth order harmonic components of voltage of α axis
$V_{\beta m}$ = Amplitude of mth order harmonic components of voltage of β axis
V_{d1} = Amplitude of fundamental components of voltage of d axis
V_{q1} = Amplitude of fundamental components of voltage of q axis
V_{dm} = Amplitude of mth order harmonic components of voltage of d axis
V_{qm} = Amplitude of mth order harmonic components of voltage of q axis
V_{RY} = Amplitude of R-phase to Y-phase voltage
V_{YB} = Amplitude of Y-phase to B-phase voltage
$[v_{R,Y,B}]$ = Voltage matrix consisting of R, Y and B phase voltages

$[v_{\alpha,\beta,0}]$ = Voltage matrix in Clarke plane
$[v_{d,q,0}]$ = Voltage matrix in Park plane
$\begin{bmatrix} v_\alpha \\ v_\beta \end{bmatrix}$ = Voltage matrix in Clarke planes
$\begin{bmatrix} v_d \\ v_q \end{bmatrix}$ = Voltage matrix in Park planes
I_R = Amplitude of current in phase R
I_Y = Amplitude of current in phase Y
I_B = Amplitude of current in phase B
I_{R0} = Zero sequence current in phase R
I_{Y0} = Zero sequence current in phase Y
I_{B0} = Zero sequence current in phase B
$\mathrm{I_{R1}}$ = Positive sequence current in phase R
$\mathrm{I_{Y1}}$ = Positive sequence current in phase Y
$\mathrm{I_{B1}}$ = Positive sequence current in phase B
$\mathrm{I_{R2}}$ = Negative sequence current in phase R
$\mathrm{I_{Y2}}$ = Negative sequence current in phase Y
$\mathrm{I_{B2}}$ = Negative sequence current in phase B
$i(t)$ = Current
$i_R(t)$ = Voltage of phase R
$i_Y(t)$ = Voltage of phase Y
$i_B(t)$ = Voltage of phase B
$i_R^N(t)$ = Normalized current in phase R
$i_Y^N(t)$ = Normalized current in phase Y
$i_B^N(t)$ = Normalized current in phase B
i_{R1} = Fundamental component of R phase current
I_{Rn} = Amplitude of nth order harmonic components of R phase current
I_{Yn} = Amplitude of nth order harmonic components of Y phase current
I_{Bn} = Amplitude of nth order harmonic components of B phase current
$I_{\alpha 1}$ = Amplitude of fundamental components of current of α axis
$I_{\beta 1}$ = Amplitude of fundamental components of current of β axis
$I_{\alpha n}$ = Amplitude of nth order harmonic components of current of α axis
$I_{\beta n}$ = Amplitude of nth order harmonic components of current of β axis
I_{d1} = Amplitude of fundamental components of current of d axis
I_{q1} = Amplitude of fundamental components of current of q axis
I_{dn} = Amplitude of nth order harmonic components of current of d axis
I_{qn} = Amplitude of nth order harmonic components of current of q axis
$\mathrm{I_L}$ = Line current
$\begin{bmatrix} i_\alpha \\ i_\beta \end{bmatrix}$ = Current matrix in Clarke planes
$\begin{bmatrix} i_d \\ i_q \end{bmatrix}$ = Current matrix in Park planes
$[i_{R,Y,B}]$ = Current matrix consisting of R, Y and B phase currents
$[i_{\alpha,\beta,0}]$ = Current matrix in Clarke plane
$[i_{d,q,0}]$ = Current matrix in Park plane

$v_{REF1}(t)$ = Reference signal for assessment of fundamental component of voltage waveform
$v_{REFm}(t)$ = Reference signal for harmonic component of voltage signal
$i_{REF1}(t)$ = Reference signal for fundamental component of current signal
$i_{REFn}(t)$ = Reference signal for harmonic component of current signal
θ = Phase difference between two phase-currents in unbalance condition
θ_C = Angular difference between two consecutive cleavages
θ_R = Resultant shift angle of current in phase R
θ_Y = Resultant shift angle of current in phase Y
θ_B = Resultant shift angle of current in phase B
Θ = Phase difference between two phase-currents at balance condition
θ_n = Phase angle of nth order harmonic component of current
θ_{R1} = Angle of fundamental component of R phase current
θ_{Rn} = Phase angle of nth order harmonic component of R phase current
θ_{Yn} = Phase angle of nth order harmonic component of Y phase current
θ_{Bn} = Phase angle of nth order harmonic component of B phase current
$\theta_{\alpha 1}$ = Phase angle of fundamental component of current of α axis
$\theta_{\beta 1}$ = Phase angle of fundamental component of current of β axis
$\theta_{\alpha n}$ = Phase angle of nth order harmonic component of current of α axis
$\theta_{\beta n}$ = Phase angle of nth order harmonic component of current of β axis
θ_{d1} = Phase angle of fundamental component of current of d axis
θ_{q1} = Phase angle of fundamental component of current of q axis
θ_{qn} = Phase angle of nth order harmonic component of current of q axis
θ_{dn} = Phase angle of nth order harmonic component of current of d axis
φ_{R1} = Angle of fundamental component of R phase voltage
φ_{Rm} = Angle of harmonic component of R phase voltage
φ_n = Phase angle n^{th} order harmonic component of voltage
ϕ_{Rm} = Phase angle of mth order harmonic component of R phase voltage
ϕ_{Ym} = Phase angle of mth order harmonic component of Y phase voltage
ϕ_{Bm} = Phase angle of mth order harmonic component of B phase voltage
$\varphi_{\alpha 1}$ = Phase angle of fundamental component of voltage of α axis
$\varphi_{\beta 1}$ = Phase angle of fundamental component of voltage of β axis
$\phi_{\alpha m}$ = Phase angle of mth order harmonic component of voltage of α axis
$\varphi_{\beta m}$ = Phase angle of mth order harmonic component of voltage of β axis
ϕ_{d1} = Phase angle of fundamental component of voltage of d axis
φ_{q1} = Phase angle of fundamental component of voltage of q axis
φ_{qm} = Phase angle of mth order harmonic component of voltage of q axis
φ_{dm} = Phase angle of mth order harmonic component of voltage of d axis
n = Order of harmonics in general
n_V = Highest order of harmonics present in voltage waveform
n_I = Highest order of harmonics present in current waveform
n_H = Order of highest harmonic
X_{MIN} = Minimum value of X
X_{MAX} = Maximum value of X

X_1 = X when Y is minimum
X_2 = X when Y is maximum
X_0 = Modulus of X when Y is zero
x = Difference of X_1 and X_2
x = Amplitude of harmonic as percentage of fundamental
$[x_V]$ = Column Matrix formed by X parameters in voltage-voltage plane
$[x_I]$ = Column Matrix formed by X parameters in current-current plane
Y_{MIN} = Minimum value of Y
Y_{MAX} = Maximum value of Y
Y_1 = Y when X is minimum
Y_2 = Y when X is maximum
Y_0 = Modulus of Y when X is zero
y = Difference of Y_1 and Y_2
$[y_V]$ = Column Matrix formed by Y parameters in voltage-voltage plane
$[y_I]$ = Column Matrix formed by Y parameters in current-current plane
ω = Fundamental angular frequency
k_1 = Constant having magnitude 0.1ω
K = Constant which depends on angular frequency
α_n = Specific angle for harmonic component
D = Depth (D) of a cleavage
P = Percent of the fundamental component.
$[Clarke\ Matrix\ or\ CM]$ = Clarke transformation matrix
$[Park\ Matrix\ or\ PM]$ = Park transformation matrix
A^{v-i}_{TOTAL} = Area formed in v–i plane
A^{vi-t}_{TOTAL} = Area under the vi–t curve
$A^{(i_R-i_{REF})}_{R1}$ = Area in $(i_R–i_{REF})$ plane contributed by fundamental current components
$A^{(i_{REF}i_R-t)}_{R1}$ = Area in $(i_{REF}i_R–t)$ plane contributed by fundamental current components of phase R
$A^{(v_{REF}v_R-t)}_{R1}$ = Area in $(v_{REF}v_R–t)$ plane contributed by fundamental voltage components of phase R
$A^{(v_R-v_{REF})}_{R1}$ = Area in $(v_R–v_{REF})$ plane contributed by fundamental component of voltage of phase R
$A^{(v_R-v_{REF})}_{Rm}$ = Area in $(v_R–v_{REF})$ plane contributed by m^{th} order voltage harmonics of phase R
$A^{(v_\alpha-v_{REF})}_{\alpha 1}$ = Area in $(v_\alpha–v_{REF})$ plane contributed by fundamental component of voltage of α axis
$A^{(v_\beta-v_{REF})}_{\beta 1}$ = Area in $(v_\beta–v_{REF})$ plane contributed by fundamental component of voltage of β axis

$A_{\alpha m}^{(v_\alpha - v_{REF})}$ = Area in $(v_\alpha - v_{REF})$ plane contributed by m[th] order voltage harmonics of α axis

$A_{\beta m}^{(v_\beta - v_{REF})}$ = Area in $(v_\beta - v_{REF})$ plane contributed by m[th] order voltage harmonics of β axis

$A_{d1}^{(v_d - v_{REF})}$ = Area in $(v_d - v_{REF})$ plane contributed by fundamental component of voltage of d axis

$A_{q1}^{(v_q - v_{REF})}$ = Area in $(v_q - v_{REF})$ plane contributed by fundamental component of voltage of q axis

$A_{dm}^{(v_d - v_{REF})}$ = Area in $(v_d - v_{REF})$ plane contributed by m[th] order voltage harmonics of d axis

$A_{qm}^{(v_q - v_{REF})}$ = Area in $(v_q - v_{REF})$ plane contributed by m[th] order voltage harmonics of q axis

$A_{R1}^{(v_{REF} v_R - t)}$ = Area in $(v_{REF} v_R - t)$ plane contributed by fundamental voltage components of phase R

$A_{Rm}^{(v_{REF} v_R - t)}$ = Area in $(v_{REF} v_R - t)$ plane contributed by m[th] order harmonic component of phase R

$A_{\alpha 1}^{(v_{REF} v_\alpha - t)}$ = Area in $(v_{REF} v_\alpha - t)$ plane contributed by fundamental voltage components of α axis

$A_{\beta 1}^{(v_{REF} v_\beta - t)}$ = Area in $(v_{REF} v_\beta - t)$ plane contributed by fundamental voltage components of β axis

$A_{\alpha m}^{(v_{REF} v_\alpha - t)}$ = Area in $(v_{REF} v_\alpha - t)$ plane contributed by m[th] order harmonic component of α axis

$A_{\beta m}^{(v_{REF} v_\beta - t)}$ = Area in $(v_{REF} v_\beta - t)$ plane contributed by m[th] order harmonic component of β axis

$A_{d1}^{(v_{REF} v_d - t)}$ = Area in $(v_{REF} v_d - t)$ plane contributed by fundamental voltage components of d axis

$A_{q1}^{(v_{REF} v_q - t)}$ = Area in $(v_{REF} v_q - t)$ plane contributed by fundamental voltage components of q axis

$A_{dm}^{(v_{REF} v_d - t)}$ = Area in $(v_{REF} v_d - t)$ plane contributed by m[th] order harmonic component of d axis

$A_{qm}^{(v_{REF} v_q - t)}$ = Area in $(v_{REF} v_q - t)$ plane contributed by m[th] order harmonic component of q axis

$A_{R1}^{(i_R - i_{REF})}$ = Area in $(i_R - i_{REF})$ plane contributed by fundamental component of current of phase R

$A_{Rn}^{(i_R - i_{REF})}$ = Area in $(i_R - i_{REF})$ plane contributed by n[th] order harmonic component of current of phase R

$A_{\alpha 1}^{(i_\alpha - i_{REF})}$ = Area in $(i_\alpha - i_{REF})$ plane contributed by fundamental component of current of α axis

$A_{\beta 1}^{(i_\beta - i_{REF})}$ = Area in $(i_\beta - i_{REF})$ plane contributed by fundamental component of current of β axis

$A_{\alpha n}^{(i_\alpha - i_{REF})}$ = Area in $(i_\alpha - i_{REF})$ plane contributed by n[th] order harmonic component of current of α axis

$A_{\beta n}^{(i_\beta - i_{REF})}$ = Area in $(i_\beta - i_{REF})$ plane contributed by n^{th} order harmonic component of current of β axis

$A_{d1}^{(i_d - i_{REF})}$ = Area in $(i_d - i_{REF})$ plane contributed by fundamental component of current of d axis

$A_{q1}^{(i_q - i_{REF})}$ = Area in $(i_q - i_{REF})$ plane contributed by fundamental component of current of q axis

$A_{dn}^{(i_d - i_{REF})}$ = Area in $(i_d - i_{REF})$ plane contributed by n^{th} order harmonic component of current of d axis

$A_{qn}^{(i_q - i_{REF})}$ = Area in $(i_q - i_{REF})$ plane contributed by n^{th} order harmonic component of current of q axis

$A_{R1}^{(i_{REF} i_R - t)}$ = Area in $(i_{REF} i_R - t)$ plane contributed by fundamental component of current of phase R

$A_{Rn}^{(i_{REF} i_R - t)}$ = Area in $(i_{REF} i_R - t)$ plane contributed by n^{th} order harmonic component of current of phase R

$A_{\alpha 1}^{(i_{REF} i_\alpha - t)}$ = Area in $(i_{REF} i_\alpha - t)$ plane contributed by fundamental component of current of α axis

$A_{\beta 1}^{(i_{REF} i_\beta - t)}$ = Area in $(i_{REF} i_\beta - t)$ plane contributed by fundamental component of current of β axis

$A_{\alpha n}^{(i_{REF} i_\alpha - t)}$ = Area in $(i_{REF} i_\alpha - t)$ plane contributed by n^{th} order harmonic component of current of α axis

$A_{\beta n}^{(i_{REF} i_\beta - t)}$ = Area in $(i_{REF} i_\beta - t)$ plane contributed by n^{th} order harmonic component of current of β axis

$A_{d1}^{(i_{REF} i_d - t)}$ = Area in $(i_{REF} i_d - t)$ plane contributed by fundamental component of current of d axis

$A_{q1}^{(i_{REF} i_q - t)}$ = Area in $(i_{REF} i_q - t)$ plane contributed by fundamental component of current of q axis

$A_{dn}^{(i_{REF} i_d - t)}$ = Area in $(i_{REF} i_d - t)$ plane contributed by n^{th} order harmonic component of current of d axis

$A_{qn}^{(i_{REF} i_q - t)}$ = Area in $(i_{REF} i_q - t)$ plane contributed by n^{th} order harmonic component of current of q axis

A_E = Area enclosed by voltage and current in one cycle in v-i plane

A_{RY} = Area formed by $V_{RY} - I_L$ curve in v-i plane

A_{YB} = Area formed by $V_{YB} - I_L$ curve in v-i plane

A_T = Total area

Q_{R1} = Reactive power contributed by fundamental components of R phase

Q_{Rn} = Reactive power contributed by harmonic component components of R Phase

$Q_{\alpha 1}$ = Reactive power contributed by fundamental components along α axis

$Q_{\beta 1}$ = Reactive power contributed by fundamental components along β axis

$Q_{\alpha n}$ = Reactive power contributed by harmonic components along α axis

$Q_{\beta n}$ = Reactive power contributed by harmonic components along β axis

Q_{d1} = Reactive power contributed by fundamental components along d axis
Q_{q1} = Reactive power contributed by fundamental components along q axis
Q_{dn} = Reactive power contributed by harmonic components along d axis
Q_{qn} = Reactive power contributed by harmonic components along q axis
P_{R1} = Active power contributed by fundamental components of R phase
P_{Rn} = Active power contributed by harmonic component components of R Phase
$P_{\alpha 1}$ = Active power contributed by fundamental components along α axis
$P_{\beta 1}$ = Active power contributed by fundamental components along β axis
$P_{\alpha n}$ = Active power contributed by harmonic components along α axis
$P_{\beta n}$ = Active power contributed by harmonic components along β axis
P_{d1} = Active power contributed by fundamental components along d axis
P_{q1} = Active power contributed by fundamental components along q axis
P_{dn} = Active power contributed by harmonic components along d axis
P_{qn} = Active power contributed by harmonic components along q axis
S_{R1} = Complex power contributed by fundamental components of R phase voltage
S_{Rn} = Complex power contributed by harmonic components of phase
S_R = Complex power contributed by fundamental and harmonic components of phase R
S_{C1} = Complex power contributed by fundamental components in Clarke plane
S_{Cn} = Complex power contributed by harmonic components in Clarke plane
S_C = Complex power contributed by fundamental and harmonic components in Clarke plane
S_{P1} = Complex power contributed by fundamental components in Park plane
S_{Pn} = Complex power contributed by harmonic components in Park plane
S_P = Complex power contributed by fundamental and harmonic components in Park plane
PDF = Active power distortion factor
QDF = Reactive power distortion factor
THD_{RV} = Total harmonic distortion of voltage in phase R
THD_{RI} = Total harmonic distortion of current in phase R
THD_{V_α} = Total harmonic distortion of α axis voltage
THD_{V_β} = Total harmonic distortion of β axis voltage
THD_{I_α} = Total harmonic distortion of α axis current
THD_{I_β} = Total harmonic distortion of β axis current
THD_{V_d} = Total harmonic distortion of d axis voltage
THD_{V_q} = Total harmonic distortion of q axis voltage
THD_{I_d} = Total harmonic distortion of d axis current
THD_{I_q} = Total harmonic distortion of q axis current
PDF_R = Active power distortion factor in phase R
PDF_α = Active power distortion factor of α axis voltage
PDF_β = Active power distortion factor of β axis voltage
PDF_d = Active power distortion factor of d axis current
PDF_q = Active power distortion factor of q axis current

QDF_R = Reactive power distortion factor in phase R
QDF_α = Reactive power distortion factor of α axis voltage
QDF_β = Reactive power distortion factor of β axis voltage
QDF_d = Reactive power distortion factor of d axis current
QDF_q = Reactive power distortion factor of q axis current
ADF_R = Apparent power distortion factor in phase R
ADF_α = Apparent power distortion factor of α axis voltage
ADF_β = Apparent power distortion factor of β axis voltage
ADF_d = Apparent power distortion factor of d axis current
ADF_q = Apparent power distortion factor of q axis current
EPQ = Electric power quality
PBM = Passivity based model
ABM = Activity based model

Chapter 1
Introduction

Abstract Electrical power quality is one of the most modern branches in power system study. This chapter starts with short definition of electric power quality. It describes in brief the causes of poor power quality in power system. Need of research on electric power quality is highlighted. At last, the content of the book at a glance is presented.

1.1 Definition of Electric Power Quality

Electric Power Quality (EPQ) is a term that refers to maintaining the near sinusoidal waveform of power distribution bus voltages and currents at rated magnitude and frequency.

1.2 Sources for Electric Power Quality Deterioration in a Power System

The sources of poor power quality can be categorized in two groups: (1) actual loads, equipment and components and (2) subsystems of transmission and distribution systems. Poor quality is normally caused by power line disturbances such as impulses, notches, voltage sag and swell, voltage and current unbalances, momentary interruption and harmonic distortions. The International Electro-technical Commission (IEC) classification of power quality includes loss-of-balance as a source of disturbance. IEEE standard also includes this feature as a source of quality deterioration of electric power. The other major contributors to poor power quality are harmonics and reactive power. Solid state control of ac power using high speed switches are the main source of harmonics whereas different non-linear loads contribute to excessive drawl of reactive power from supply. It leads to catastrophic consequences such as long production downtimes, mal-function of devices and shortened equipment life.

S. Chattopadhyay et al., *Electric Power Quality,* Power Systems,
DOI 10.1007/978-94-007-0635-4_1, © Springer Science+Business Media B.V. 2011

1.3 Need for Assessment of Electric Power Quality

It is common experience that electric power of poor quality has detrimental effects on health of different equipment and systems. Moreover, power system stability, continuity and reliability fall with the degradation of quality of electric power. For example, it has been reported that a 10% increase in voltage stress caused by harmonic currents typically results in 7% increase in the operating temperature of a capacitor bank and can reduce its life expectancy by 30% of normal. To avoid such effects, it is thus of utmost importance to continuously assess the quality of power supplied to a consumer. Moreover present day deregulated scenario of power network demands high quality electric power.

1.4 Book at a Glance

After giving a short introduction in this chapter, Chap. 2 deals with electric power quality in power system. It describes what is quality of power and main causes and effects of poor power quality. Different power quality related IEC and IEEE standards are mentioned.

Chapter 3 deals with unbalance with its main causes and effects in power system.

Harmonic is an important power system disturbance. Details of harmonics along with definition, sources and effects of harmonics are discussed in Chap. 4.

Transient is another power quality related disturbance. Types, sources and effects of transients are discussed in Chap. 5.

Sag, swell, interruption, under voltage and over voltage are discussed in Chap. 6.

DC offset, ringing wave, flicker, etc. are discussed in Chap. 7.

Assessment of main power quality disturbance starts from Chap. 8. In this chapter, unbalance is assessed using sequence components.

Chapter 9 presents feature pattern extraction method (FPEM) for monitoring EPQ of a power system in respect of unbalance.

Different existing useful tools for harmonic assessment are mentioned in Chap. 10.

In Chap. 11, feature pattern extraction method is used for harmonics assessment in voltage–voltage and current–current plane. Merits and demerits of this method for harmonic assessment are discussed in this chapter.

In Chap. 12, fundamentals of Clarke and Park transformation as used in three-phase analysis are described.

In Chap. 13, attempt has been made to overcome the limitations of harmonic assessment in voltage–voltage and current–current plane using feature pattern extraction method. The method is used for harmonic assessment in Clarke and Park plane.

In Chap. 14, area based technique discussed and used in harmonic assessment.

Chapter 15 applies area based approach for assessment of harmonic distortion in Clarke and Park planes for a three-phase power system.

In the first part of Chap. 16, different power components are assessed by feature pattern extraction method and area based technique.

Chapter 17 deals with some techniques used for power system transient assessment.

Chapter 18 deals with the modeling of a poly phase system in presence of harmonics.

Main features of this edition are as follows

- Unbalance has been defined and assessed with respect to phase angle shift of the voltages and currents.
- Feature pattern extraction method has been introduced in voltage–voltage and current–current planes.
- Rule set has been developed and unbalance has been assessed using the rule set.
- Rule set has been developed for assessment of highest order of harmonics in voltage–voltage and current–current planes.
- Rule set has been developed for harmonic assessment using feature pattern extraction method in Clarke and Park planes.
- Area based approach has been made for harmonic assessment. Mathematical formulas have been derived for total harmonic distortion. Complete harmonic assessment has been done by this approach.
- Different power related parameters have been assessed using feature pattern extraction method.
- Powers contributed by fundamental as well as harmonic components of the voltages and current have been assessed separately by area based approach.
- Distortion factors with respect to different powers have been formulated and assessed by area based approach.
- Passivity based and activity based models have been developed for a polyphase system in presence of harmonics.

Chapter 2
Electric Power Quality

Abstract The chapter starts with an introduction of power quality. Different aspects are then discussed to define electric power quality. Different sub-branches in power quality study are discussed. After this, disturbances normally occurred in power system are discussed. Short definitions of these power system disturbances are presented. Power quality related problems are summarized. Different guidelines given by IEC, IEEE, etc. are presented in tabular form.

2.1 Introduction

Development of technology in all its areas is progressing at a faster rate. Power scenario has changed a lot. With the increase of size and capacity, power systems have become complex leading to reduced reliability. But, the development of electronics, electrical device and appliances have become more and more sophisticated and they demand uninterrupted and conditioned power. These have pushed the present complex electricity network and market in a strong competition resulting in the concept of deregulation. In this ever changing power scenario, quality assurance of electric power has also been affected. It demands a deep research and study on the subject 'Electric Power Quality'.

2.2 Electric Power Quality

Electric Power Quality (EPQ) is a term that refers to maintaining the near sinusoidal waveform of power distribution bus voltages and currents at rated magnitude and frequency. Thus EPQ is often used to express voltage quality, current quality, reliability of service, quality of power supply, etc.

EPQ has captured increasing attention in power engineering in recent years. In the study of EPQ, different branches are being formed. They deal with different issues related to power quality. Study on electric power quality may be divided into following stages [1–15]:

S. Chattopadhyay et al., *Electric Power Quality*, Power Systems,

DOI 10.1007/978-94-007-0635-4_2,

1. Fundamental concepts
2. Sources
3. Effects
4. Modeling and Analysis
5. Instrumentation
6. Solutions

All branches are inter-related and very much depended on each other. 'Fundamental concept' of EPQ, identifies the parameters and their degree of variation with respect to their rated magnitude which are the base reason for degradation of quality of electric power. 'Sources' are the regions or locations or events which causes the unwanted variation of those parameters. It's really a big challenge to the power engineers to find out the exact sources of power quality related disturbance in the ever increasing complex network. 'Effects' of poor quality of power are the effects faced by the system and consumer equipment after the occurrence of different disturbances. In 'modeling and analysis', attempts are taken to configure the disturbance, its occurrence, sources and effect; mainly based on the mathematical background. For monitoring of EPQ, constant measurement and 'instrumentation' of the electric parameters are necessary. Complete solution, i.e. delivery of pure power to the consumer side is practically impossible. Our target is to minimize the probability of occurrence of disturbances and to reduce the effects of EPQ problems.

EPQ describes the variation of voltage, current and frequency in a power system. Most power system equipment has been able to operate successfully with relatively wide variations of these three parameters. However, within the last five to fifteen years, a large amount of equipment has been added to the power system, which is not so tolerant of these variations. The sophistication of electrical appliances with the development of electronics has added to the demand of quality power at the consumer premises. To ensure uninterrupted and quality power has thus become a point of competition for the power producers. Thus an open and competitive power market has paved its way. These situations have introduced the concept of deregulation in power sector. Like all other commodities, for electric power there should be quality issues at each physical location in all system especially in deregulated system.

Poor power quality sources can be divided in two groups: (1) actual loads, equipment and components and (2) subsystems of transmission and distribution systems. Quality degradation of electric power is mainly occurred due to power line disturbances such as impulses, notches, voltage sag and swell, voltage and current unbalances, momentary interruption and harmonic distortions, different standards and guidelines of which are mentioned in the International Electro-technical Commission (IEC) classification of power quality and relevant IEEE standard. The other major contributors to poor power quality are harmonics and reactive power. Solid state control of ac power using high speed switches are the main source of harmonics whereas different non-linear loads contribute to excessive drawl of reactive power from supply.

2.3 Classification of Power System Disturbances

Power quality problems occur due to various types of electrical disturbances. Most of the EPQ disturbances depend on amplitude or frequency or on both frequency and amplitude. Based on the duration of existence of EPQ disturbances, events can divided into short, medium or long type. The disturbances causing power quality degradation arising in a power system and their classification mainly include:

1. ***Interruption/under voltage/over voltage:*** these are very common type disturbances. During power interruption, voltage level of a particular bus goes down to zero. The interruption may occur for short or medium or long period. Under voltage and over voltage are fall and rise of voltage levels of a particular bus with respect to standard bus voltage. Sometimes under and over voltages of little percentage is allowable; but when they cross the limit of desired voltage level, they are treated as disturbances. Such disturbances are increasing the amount of reactive power drawn or deliver by a system, insulation problems and voltage stability.
2. ***Voltage/Current unbalance:*** voltage and current unbalance may occur due to the unbalance in drop in the generating system or transmission system and unbalanced loading. During unbalance, negative sequence components appear. T hampers system performance may change loss and in some cases it may hamper voltage stability.
3. ***Harmonics:*** harmonics are the alternating components having frequencies other than fundamental present in voltage and current signals. There are various reasons for harmonics generation like non linearity, excessive use of semiconductor based switching devices, different design constrains, etc. Harmonics have adverse effects on generation, transmission and distribution system as well as on consumer equipments also. Harmonics are classified as integer harmonics, sub harmonics and inter harmonics. Integer harmonics have frequencies which are integer multiple of fundamental frequency, sub harmonics have frequencies which are smaller than fundamental frequency and inter harmonics have frequencies which are greater than fundamental frequencies. Among these entire harmonics integer and inter harmonics are very common in power system. Occurrence of sub harmonics is comparatively smaller than others. Sometimes harmonics are classified: time harmonics and spatial (space) harmonics. Obviously their causes of occurrence are different. Harmonics are in general are not welcome and desirable. Harmonics are assessed with respect to fundamental. Monitoring of harmonics with respect to fundamental is important consideration in power system application. For this purpose different distortion factor with respect to the fundamental have been introduced.
4. ***Transients:*** transients [16, 17] may generate in the system itself or may come from the other system. Transients are classified into two categories: dc transient and ac transient. AC transients are further divided into two categories: single cycle and multiple cycles.

Table 2.1 Definition of power system disturbances

Sl No	Disturbance	Short definition
A	Interruption	voltage magnitude is zero
	Under voltage	voltage magnitude is below its nominal value
	Over voltage	voltage magnitude is above its nominal value
B	*Voltage sag*	A reduction in RMS voltage over a range of 0.1–0.9 pu for a duration greater than 10 ms but less than 1 s
C	*Voltage swell*	An increase in RMS voltage over a range of 1.1–1.8 pu for a duration greater than 10 ms but less than 1 s
D	*Flicker*	A visual effect of frequency variation of voltage in a system
E	*Voltage/Current unbalance*	Deviation in magnitude of voltage/current of any one or two of the three phases
F	*Ringing waves*	A transient condition which decays gradually
G	*Outage*	Power interruption for not exceeding 60 s duration due to fault or maltripping of switchgear/system
H	Transients	Sudden rise of signal
I	Harmonics	Non-sinusoidal wave forms

5. ***Voltage sag:*** it is a short duration disturbance [18]. During voltage sag, r. m. s. voltage falls to a very low level for short period of time.
6. ***Voltage swell:*** it is a short duration disturbance. During voltage sag, r. m. s. voltage increases to a very high level for short period of time.
7. ***Flicker:*** it is undesired variation of system frequency.
8. ***Ringing waves:*** oscillatory disturbances of decaying magnitude for short period of time is known as ringing wave. It may be called a special type transient. The frequency of a flicker may or may not be same with the system frequency.
9. ***Outage:*** it is special type of interruption where power cut has occurred for not more than 60 s.

Short definitions of the power system disturbances are summarized in Table 2.1 [16–30].

2.4 Power Quality Standards and Guidelines

Standards and guidelines have been given by different technical bodies like IEEE, ANSI, IEC, etc. Those guidelines are very helpful in EPQ study and practice. Some references related to EPQ with their main content are presented in Tables 2.2 and 2.3 [31–37].

Table 2.2 IEEE and ANSI guidelines

IEEE 4	Standard techniques for high-voltage testing
IEEE 100	Standard dictionary of electrical and electronic engineering
IEEE 120	Master test guide for electrical measurements in power circuits
IEEE 141	Recommended practice for electric power distribution for industrial plants with effect of voltage disturbances on equipment within an industrial area
IEEE 142	Recommended practice for grounding of industrial and commercial power systems
IEEE 213	Standard procedure for measurement of conducted emissions in the range of 300 kHz–25 MHz from television and FM broadcast receivers to power lines
IEEE 241	Recommended practice for electric power systems in commercial buildings
IEEE 281	Standard service conditions for power system communication equipment
IEEE 299	Standard methods of measuring the effectiveness of electromagnetic shielding enclosures
IEEE 367	Recommended practice for determining the electric power station ground potential rise and induced voltage from a power fault
IEEE 376	Standard for the measurement of impulse strength and impulse bandwidth
IEEE 430	Standard procedures for the measurement of radio noise from overhead power lines and substations
IEEE 446	Recommended practice for emergency and standby systems for industrial and commercial applications (e.g., power acceptability curve, CBEMA curve)
IEEE 449	Standard for ferro resonance voltage regulators
IEEE 465	Test specifications for surge protective devices
IEEE 472	Event recorders
IEEE 473	Recommended practice for an electromagnetic site survey (10 kHz–10 GHz)
IEEE 493	Recommended practice for the design of reliable industrial and commercial power systems
IEEE 519	Recommended practice for harmonic control and reactive compensation of static power converters
IEEE 539	Standard definitions of terms relating to corona and field effects of overhead power lines
IEEE 859	Standard terms for reporting and analyzing outage occurrences and outage states of electrical transmission facilities
IEEE 944	Application and testing of uninterruptible power supplies for power generating stations
IEEE 998	Guides for direct lightning strike shielding of substations
IEEE 1048	Guides for protective grounding of power lines
IEEE 1057	Standards for digitizing waveform recorders
IEEE Pl100	Recommended practice for powering and grounding sensitive electronic equipment in commercial and industrial power systems
IEEE 1159	Recommended practice on monitoring electric power quality. Categories of power system electromagnetic phenomena
IEEE 1250	Guides for service to equipment sensitive to momentary voltage disturbances
IEEE 1346	Recommended practice for evaluating electric power system compatibility with electronics process equipment
IEEE P1453	Flicker

Table 2.2 (continued)

IEEE/ANSI 18	Standards for shunt power capacitors
IEEE/ANSI C37	Guides for surge withstand capability (SWC) tests
IEEE/ANSI C50	Harmonics and noise from synchronous machines
IEEE/ANSI C57.110	Recommended practice for establishing transformer capability when supplying no sinusoidal load currents
IEEE/ANSI C57.117	Guides for reporting failure data for power transformers and shunt reactors on electric utility power systems
IEEE/ANSI C62.45 (IEEE 587)	Recommended practice on surge voltage in low-voltage AC power circuits, including guides for lightning arresters applications
IEEE/ANSI C62.48	Guides on interactions between power system disturbances and surge protective devices
ANSI C84.1	American national standard for electric power systems and equipment voltage ratings (60 Hz)
ANSI 70	National electric code
ANSI 368	Telephone influence factor
ANSI 377	Spurious radio frequency emission from mobile communication equipment

Table 2.3 IEC guidelines

IEC 38	Standard voltages
IEC 816	Guides on methods of measurement of short-duration transients on low-voltage power and signal lines. Equipment susceptible to transients
IEC 868	Flicker meter. Functional and design specifications
IEC 868-0	Flicker meter. Evaluation of flicker severity. Evaluates the severity of voltage fluctuation on the light flicker
IEC 1000-3-2	Electromagnetic compatibility Part 3: Limits Section 2: Limits for harmonic current emissions (equipment absorbed current <16 A per phase)
IEC 1000-3-6	Electromagnetic compatibility Part 3: Limits Section 6: Emission limits evaluation for perturbing loads connected to MV and HV networks
IEC 1000-4	Electromagnetic compatibility Part 4: Sampling and metering techniques
EN 50160	Voltage characteristics of electricity supplied by public distribution systems
EC/EN 60868	Flicker meter implementation
IEC 61000	Electromagnetic compatibility (EMC)

References

[1] Sankaran, C.: Power Quality. CRC Press, Boca Raton (2002)

[2] Gosbell, V.J., Perera, B.S.P., Herath, H.M.S.C.: New framework for utility power quality (PQ) data analysis. Proceedings AUPEC'01, Perth, pp. 577–582 (2001)

[3] Bollen, M.H.J.: Understanding Power Quality Problems-Voltage Sags and Interruptions. IEEE Press, New York (2001)

[4] Arrillaga, J., Watson, N.R., Chen, S.: Power System Quality Assessment. Wiley, New York (2000)

[5] Shaw, S.R., Laughman, C.R., Leeb, S.B., Lepard, R.F.: A power quality prediction system. IEEE Trans. Ind. Electron. **47**(3), 511–517 (2000)

[6] Santoso, S., Lamoree, J., Grady, W.M., Powers, E.J., Bhatt, S.C.: A scalable PQ event identification system. IEEE Trans. Power Deliv. **15**, 738–742 (2000)

[7] Santoso, S., Powers, E.J., Grady, W.M., Parsons, A.C.: Characterization of distribution power quality events with Fourier and wavelet transform. IEEE Trans. Power Deliv. **15**(1), 247–256 (2000)

[8] Watson, N.R., Ying, C.K., Arnold, C.P.: A global power quality index for aperiodic waveforms. Proceedings IEEE 9th International Conference on Harmonies and Quality of Power, pp. 1029–1034 (2000)
[9] Domijan, A., Heydt, G.T., Mellopouloe, A.P.S., Venkata, S.S., West, S.: Directions of research on power quality. IEEE Trans. Power Deliv. **8**(1), 429–436 (1993)
[10] IEEE Standard 1195: IEEE recommended practices for monitoring power quality, pp. 1–59. IEEE Inc., New York (1995)
[11] IEEE Standard 519: IEEE recommended practices and requirements for harmonic control in electric power systems. IEEE-519, 1992. Standard power systems, IEEE-519 (1992)
[12] IEEE Working Group: Power quality-two different perspective. IEEE Trans. Power Deliv. **5**(3), 1501–1513 (1990)
[13] Duffey, C.K., Stratford, R.P.: Update of harmonic standard IEEE-519: IEEE recommended practices and requirements for harmonic control in power systems. IEEE Trans. Ind. Appl. **25**(6), 1025–1034 (1989)
[14] Fuller, J.F., Fuchs, E.F., Roesler, D.J.: Influence of harmonics on power system distribution protection. IEEE Trans. Power Deliv. **TPWRD-3**(2), 546–554 (1988)
[15] Fuchs, E.F., Roesler, D.J., Kovacs, K.P.: Aging of electrical appliances due to harmonics of the power system's voltage. IEEE Trans. Power Deliv. **TPWRD-1**(3), 301–307 (1986)
[16] Bollen, M.H.J., Styvaktakis, E., Yu-HuaGu, I.: Categorization and analysis of power system transients. IEEE Trans. Power Deliv. **20**(3), 2298–2306 (2005)
[17] Herath, C., Gosbell, V., Perera, S.: A transient index for reporting power quality (PQ) surveys. Proceedings CIRED 2003, pp. 2.61-1–2.61-5. Bercelona, Spain (2003)
[18] Djokic, S.Z., Desmet, J., Vanalme, G., Milanovic, J.V., Stockman, K.: Sensitivity of personal computer to voltage sags and short interruption. IEEE Trans. Power Deliv. **20**(1), 375–383 (2005)
[19] Lin, D., Fuchs, E.F.: Real-time monitoring of iron-core and copper losses of three-phase transformer under (non)sinusoidal operation. IEEE Trans. Power Deliv. **21**(3), 1333–1341 (2006)
[20] Herath, H.M.S.C., Gosbell, V.J., Perera, S.: Power quality (PQ) survey reporting: Discrete Disturbance limit. IEEE Trans. Power Deliv. **20**(2), 851–858 (2005)
[21] Beaulieu, G., Bollen, H.J.M., Koch, R.G., Malgaroti, S., Mamo, X., Sinclair, J.: Power quality indices and objectives for MV, HV, and EHV systems CIGRE WG 36.07/CIRED progresses. Proceedings CIRED 2003, pp. 2.74-1–2.74-6. Bercelona, Spain (2003)
[22] Styvaktakis, E., Bollen, M.H.J., Yu-HuaGu, I.: Expert system for classification and analysis of power system events. IEEE Trans. Power Deliv. **17**(2), 423–428 (2002)
[23] Huang, J., Negnevitsky, M., Thong Nguyen, D.: A neural-fuzzy classifier for recognition of power quality disturbances. IEEE Trans. Power Deliv. **17**(2), 609–616 (2002)
[24] Francois, D.M., Thomas, M.G.: Power quality site surveys: Facts, fiction, and fallacies. IEEE Trans. Ind. Appl. **24**(6) (1998)
[25] Ronald, H.S.: Instrumentation, measurement techniques and analytical tools in power quality studies. Proceedings IEEE, Annual Conference of Pulp and Paeu Industry, pp. 119–140 (1997)
[26] IEEE Working Group on Non-sinusoidal Situations: Practical definitions for powers in system with non-sinu wave forms, unbalanced cond: A discussion. IEEE Trans. Power Deliv. **II**, 79–101 (1996)
[27] Cristaldi, L., Ferrero, A.: Harmonic power flow analysis for the measurement of the electric power quality. IEEE Trans. Instrum. Meas. **44**(3), 683–685 (1995)
[28] IEEE Recommended Practice for Monitoring Electric Power Quality, IEEE Standard 1159-1995 (1995)
[29] Barker, P.P., Short, T.A., Burns, C.W., Burke, J.J., Warren, C.A., Siewierski, J.J., Mancao, R.T.: Power quality monitoring of a distribution system. IEEE Trans. Power Deliv. **9**(2), 429–436 (1994)
[30] Douglas, J.: Power quality solutions. IEEE Power Eng. Rev. **14**(3), 3–7 (1994)
[31] Bollen, M.H.J.: Understanding power quality problems. IEEE Press Ser Power Eng. (2000)
[32] Standard ANSI C84.1

[33] Standard IEEE-1159
[34] Standard EN-50160
[35] Standard IEEE-1250
[36] Dugan, R.C., McGranaghan, M.F., Beaty, H.W.: Electrical Power Systems Quality. McGraw-Hill, New York (1996)
[37] Ewald, F.F., Mahammad, A., Masoum, S.: Power Quality in Power Systems and Electrical Machines, AP, ISBN 978-0-12-369536

Chapter 3
Unbalance

Abstract The chapter starts with an introduction of unbalance. Conventional definition of unbalance is discussed. It then describes main sources of unbalance in electric power system. Then effects of unbalance in power system are mentioned.

3.1 Introduction

Unbalance is a common type of power quality problem. It refers to the deviation of phase voltages and phase currents from their rated values with respect to magnitude and phase. Sequence components occur during unbalance in three phase system. For this reason unbalance is assessed with respect to the sequence components present in the system. Exact unbalance measurement without phase angle assessment is possible [1]. In some cases unbalance in presence of harmonics and inter harmonics is important [2]. Characterization of unbalance in power system is done using symmetrical components [3–5].

3.2 Unbalance in Three Phase Power System

Unbalance in power system is defined as deviation in magnitude of voltage/current of any one or two of the three phases. When voltages of a three-phase system are not identical in magnitude and/or the phase differences between them are not exactly 120 °, voltage unbalance occurs. It is often called as voltage unbalance. There are two ways of calculating the degree of unbalance:

- dividing the maximum deviation from the average of three-phase voltages by the average of three phase voltages, or
- computing the ratio of the negative (or zero) sequence component to the positive-sequence component.

Thus, unbalance in power system can be expressed as the percentage change in line currents or voltages from rated values. For change in line current in any phase among

S. Chattopadhyay et al., *Electric Power Quality,* Power Systems,
DOI 10.1007/978-94-007-0635-4_3,

the three phases R, Y and B, unbalance will be

$$\frac{I - i}{I} \times 100\% \tag{3.1}$$

where, I = rated current and i = actual current

Similarly, in respect of voltage, unbalance can be written as

$$\frac{V - v}{V} \times 100\% \tag{3.2}$$

where, V = rated voltage and v = actual voltage

Unbalance in power system is also characterized with the help of symmetrical components. "True" unbalance factor UF is defined as

$$UF = \frac{V_-}{V_+} \tag{3.3}$$

V_+ and V_- represent the root mean square (RMS) voltages of the positive and negative sequence components, respectively. Line voltage drop due to unbalance has been formulated for some events in power system. A lot of research work has been done to assess the unbalance. Some of them have not measured phase angles.

3.3 Sources of Unbalance

Unbalance may generate due to the unequal drops in individual phase or unbalanced in three phase loading. Even it may happen due the source, load, improper grounding, etc. The main causes of voltage due to unbalance in power systems are

- Unbalanced single-phase loading in a three-phase system: most of the domestic loads and industrial lighting loads are single phase. However, these loads are fed from three phase supply. If the load divisions among different phases are not coordinated, the phase parameters may differ from each other causing unbalanced demand from the supply.
- Overhead transmission lines that are not transposed,
- Blown out fuses in one phase of a three-phase capacitor bank, and
- Severe voltage unbalance (e.g., >5%), which can result from single phasing conditions.

3.4 Effect of Unbalance

Unbalance in power system is related to the power system stability problem. Unbalance may cause excessive drawl of reactive power, mal-operation of equipment, mal-operation of measuring instruments and shortening of life span of different appliances. Reactive power compensation required for individual phase will differ from each other. Performance of FACTs controller degrades during voltage unbalance. During current unbalance negative sequence component appears. It increases net current in some phase and decreases net current in other phase. This results in unequal loss in phases and unequal heating. Three phase motors draw unbalanced current from unbalanced supply system. In such situation, unequal heating and oscillation in torque hamper motors' performance. In ungrounded system, unbalance causes neutral shifting. This hampers accurate operation of relays and circuit breakers along with other related problems.

References

[1] Ghijselen, J.A.L., Van den Bossche, A.P.M.: Exact voltage unbalance assessment without phase measurement. IEEE Trans. Power Syst. **20**(1), 519–520 (2005)

[2] Leva, S., Moronado, A.P., Zaninelli, D.: Evaluation of the line voltage drops in presence of unbalance, harmonics and inter-harmonics: Theory and applications. IEEE Trans. Power Deliv. **20**(1), 390–396 (2005)

[3] Bollen, M.: Definitions of symmetrical coordinates applied to the solution of poly phase networks. IEEE Power Eng. Rev. **22**(11), 49–50 (2002)

[4] Bollen, M.H.J., Styaktakis, E.: Characterization of three phase unbalance dips (as easy as one-two-three?). Proceedings 9th International IEEE Conference Harmonics Quality Power, vol. I, Orlando, pp. 81–86, 1–4 Oct 2000

[5] Zhang, L.D., Bollen, M.H.J.: A method for characterizing unbalanced voltage dips (sags) with symmetrical components. IEEE Power Eng. Rev. **18**, 50–52 (1998)

Chapter 4
Harmonics

Abstract The chapter starts with an introduction of power system harmonics. It defines integer and non-integer types of harmonics. Then it describes the main sources of harmonics. After this, effects of harmonics in electrical machines and power system are discussed. Harmonic standards given by IEC and IEEE are referred. At the end different types of harmonic related indices are presented based on which power quality assessment may be done.

4.1 Introduction

A pure polyphase system is expected to have pure sinusoidal alternating current and voltage waveforms of single frequency. But, the real situation deviates from this purity. Real voltage and current waveforms are distorted. Normally they are called non sinusoidal waveforms. Non-sinusoidal waveform is formed with the combination of many sine waves of different frequencies. Thus actual power system signals have fundamental component as well as harmonic components.

4.2 Fundamental Wave

The waveform in a non-sinusoidal wave having frequency at which the system is designed and is expected to run and/or is being operated is known as fundamental wave. The corresponding frequency is called fundamental frequency. Fundamental wave is a pure sine wave having fundamental frequency. In some part of the world like India fundamental frequency is 50 Hz and other existing value is 60 Hz. Let 'f' be the fundamental frequency in current waveforms, then fundamental component of current will be

$$i_1 = I_1 \sin 2\pi ft \tag{4.1}$$

where, I_1 is amplitude of fundamental component of current.

S. Chattopadhyay et al., *Electric Power Quality*, Power Systems,
DOI 10.1007/978-94-007-0635-4_4, © Springer Science+Business Media B.V. 2011

4.3 Harmonics

Waveforms in a non-sinusoidal wave having frequencies other than fundamental frequency are called harmonics. The frequency of each harmonic component is known as harmonic frequency. In most of the cases of periodic and well defined waves, where waveform can be expressed by Fourier series, harmonic frequencies are integer multiple of fundamental frequency (however it may be fractional). The integer factor is known as the order of harmonic component. Let 'n' be the order, then n will be 1, 2, 3, ..., etc. Harmonic component of current of order n can be represented as

$$i_n = I_n \sin 2\pi nft \tag{4.2}$$

where I_n is the amplitude of harmonic component of order n. Integer harmonics are divided into two categories: odd harmonics and even harmonics. Other than integer harmonics there are sub and inter harmonics where n is fractional.

Odd Harmonics Integer harmonics having frequencies which are odd integer multiple of fundamental frequency are known as odd harmonics. Odd harmonics may be expressed as

$$i_n = I_n \sin 2\pi nft \tag{4.3}$$

where, n = 3, 5, 7, ..., etc. and I_n is the amplitude of harmonic component of order n.

Even Harmonics Integer harmonics having frequencies which are even integer multiple of fundamental frequency are knows as even harmonics. Even harmonics may be expressed as

$$i_n = I_n \sin 2\pi nft \tag{4.4}$$

where, n = 2, 4, 6, ..., etc. and I_n is the amplitude of harmonic component of order n.

Non-sinusoidal Waveform Non-sinusoidal wave is constituted by the combination of odd-even harmonic components as well as fundamental component. Thus, mathematically, it can be expressed as,

$$i = \sum_{n=1,2,3,4,\ldots} i_n = \sum_{n=1,2,3,4,\ldots} I_n \sin 2\pi nft \tag{4.5}$$

A non sinusoidal wave may be expressed in terms of harmonics as

$$\begin{aligned} i &= \sum_{n=1,2,3,4,\ldots} i_n = \sum_{n=1,2,3,4,\ldots} I_n \sin 2\pi nft \\ &= I_1 \sin 2\pi ft + I_2 \sin 4\pi ft + I_3 \sin 6\pi nft + I_4 \sin 8\pi nft \\ &\quad + I_5 \sin 10\pi nft + \cdots\cdots\cdots\cdots\cdots \\ &= I_1 \sin 2\pi f_1 ft + I_2 \sin 2\pi f_2 t + I_3 \sin 2\pi f_3 t + I_4 \sin 2\pi f_4 t \\ &\quad + I_5 \sin 2\pi f_5 t \ldots\ldots\ldots\ldots\ldots \end{aligned} \tag{4.6}$$

Harmonic component of order n is

$$i_n = I_n \sin 2\pi nft = I_n \sin 2\pi f_n t \tag{4.7}$$

where, $f_n = nf$

Considering phase angle, the nth order harmonic component can be expressed as

$$i_n = I_n \sin(2\pi f_n t - \alpha_n) \tag{4.8}$$

where, α_n is the phase angle of nth order current harmonic component. Considering phase angles of harmonic components, non-sinusoidal wave can be written as

$$i = \sum_{n=1,2,3,\ldots\ldots} i_n = \sum_{n=1,2,3,\ldots\ldots} I_n \sin(2\pi f_n t - \alpha_n) \tag{4.9}$$

Inter Harmonics Often in non sinusoidal waveform there are harmonics having frequencies which are greater than fundamental but not integer multiple of fundamental frequency [1]. These are known as inter-harmonics. Mathematically,

$$i_n = I_n \sin 2\pi nft \tag{4.10}$$

where, n > 1 *but not integer; e.g.*: 1.2, 1.5, 2.7 . . . *etc*

Sub Harmonic Often in non sinusoidal waveform there are harmonics having frequencies which are smaller than fundamental frequency. These are known as sub-harmonics. Mathematically,

$$i_n = I_n \sin 2\pi nft \tag{4.11}$$

where, n < 1; *e.g.*: 0.2, 0.5, 0.7 . . . *etc*

Average Value Harmonic components having phase angle (α_n) can be expressed as

$$i_n = I_n \sin(2\pi f_n t - \alpha_n)$$

Average value is given by

$$i_{nav} = \left(\frac{2}{\pi}\right)\left(\frac{I_n}{n}\right) \tag{4.12}$$

Average value of current is given as

$$i_{av} = \sum_{n=1,2,3,\ldots\ldots} \left(\frac{2}{\pi}\right)\left(\frac{I_n}{n}\right) \tag{4.13}$$

Rms Value Rms value of the non sinusoidal current wave is given by

$$i_{rms} = \sqrt{average\ of\ i^2} = \sqrt{\sum_{n=1,2,3,\ldots} \frac{I_n^2}{n}} \tag{4.14}$$

Form Factor Form factor is the ratio of rms value to average value. In case of non sinusoidal wave it is given by

$$Form\ Factor = \frac{RMS\ Value}{Average\ Value} = \frac{\sqrt{\sum_{n=1,2,3,...} \frac{I_n^2}{n}}}{\sum_{n=1,2,3,...} \left(\frac{2}{\pi}\right)\left(\frac{I_n}{n}\right)} \tag{4.15}$$

Harmonic Power Let, current waveform is

$$i = \sum_{n=1,2,3,......} i_n = \sum_{n=1,2,3,......} I_n \sin(2\pi f_n t - \alpha_n) \tag{4.16}$$

Voltage waveform is

$$v = \sum_{n=1,2,3,......} v_n = \sum_{n=1,2,3,......} V_n \sin(2\pi f_n t - \beta_n) \tag{4.17}$$

Power contributed by harmonic components of voltage and current waveforms can be expressed as

Average power of harmonics of order $n = P_n$

$$\begin{aligned}
&= \frac{1}{2\pi}\int_0^{2\pi} i_n v_n d(2\pi ft) \\
&= \frac{1}{2\pi}\int_0^{2\pi} I_n \sin(2\pi f_n t - \alpha_n) V_n \sin(2\pi f_n t - \beta_n)\, d(2\pi ft) \\
&= \frac{I_n}{\sqrt{2}}\frac{V_n}{\sqrt{2}} \cos(\alpha_n - \beta_n) \\
&= i_{n\ rms} v_{n\ rms} \cos\varphi_n
\end{aligned} \tag{4.18}$$

where,

$$\begin{aligned}
\varphi_n &= (\alpha_n - \beta_n) \\
&= \textit{phase difference between harmonic component of} \\
&\quad \textit{voltage and current waveforms of order n}
\end{aligned}$$

Total Active Power Total active power is contributed by fundamental as well as harmonic components of voltage and current waveform. Thus total power is written as

$$\begin{aligned}
p &= \sum_n P_n \\
&= \sum_n i_{n\ rms} v_{n\ rms} \cos\varphi_n \\
&= i_{1\ rms} v_{1\ rms} \cos\varphi_1 + i_{2\ rms} v_{2\ rms} \cos\varphi_2 + i_{3\ rms} v_{3\ rms} \cos\varphi_3 \\
&\quad + \cdots\cdots\cdots\cdots\cdots
\end{aligned} \tag{4.19}$$

Table 4.1 Harmonics related definitions.

Sl. No.	Parameter	Mathematical definition
1	Harmonics	$i_n = I_n \sin 2\pi nft$ n may be fractional or integer
2	Odd harmonics	$i_n = I_n \sin 2\pi nft$ n is odd integer
3	Even harmonics	$i_n = I_n \sin 2\pi nft$ n is even integer
4	Sub harmonics	$i_n = I_n \sin 2\pi nft$ n is fractional and less than 1
5	Inter harmonics	$i_n = I_n \sin 2\pi nft$ n is fractional and greater than 1
6	Average value	$i_{av} = \sum_{n=1,2,3,......} \left(\frac{2}{\pi}\right)\left(\frac{I_n}{n}\right)$
7	rms value	$i_{rms} = \sqrt{average\ of\ i^2} = \sqrt{\sum_{n=1,2,3,...} \frac{I_n^2}{n}}$
8	Form factor	$\frac{RMS\ Value}{Average\ Value} = \frac{\sqrt{\sum_{n=1,2,3,...} \frac{I_n^2}{n}}}{\sum_{n=1,2,3,...} \left(\frac{2}{\pi}\right)\left(\frac{I_n}{n}\right)}$
9	Harmonic power	$= i_{n\,rms} v_{n\,rms} \cos \varphi_n$
10	Total active power	$p = \sum_n P_n$
	Power factor	$Power\ Factor = \frac{Active\ Power}{Apparant\ Power} = \frac{\sum_n i_{n\,rms}\ v_{n\,rms} \cos \varphi_n}{i_{rms} v_{rms}}$

Power Factor Power factor of non sinusoidal waveform is written as

$$Power\ Factor = \frac{Active\ Power}{Apparant\ Power} = \frac{\sum_n i_{n\ rms}\ v_{n\ rms} \cos \varphi_n}{i_{rms} v_{rms}}$$

$$where,\ i_{rms} = \sqrt{average\ of\ i^2} = \sqrt{\sum_{n=1,2,3,...} \frac{I_n^2}{n}}$$

$$v_{rms} = \sqrt{average\ of\ v^2} = \sqrt{\sum_{n=1,2,3,...} \frac{v_n^2}{n}} \tag{4.20}$$

All the above definitions are presented in Table 4.1 [1–3].

4.4 Sources of Harmonics

Conventional electromagnetic devices as well as semiconductor applications act as sources of harmonics. Conventional electromagnetic devices include stationary transformer as well as rotating machines. Harmonic generation in these machine depends on the properties of the materials used to construct them, different design constraints and considerations, operating principle and of course load environment. Beside these arcing devices produces considerable amount of harmonics. Other than conventional devices, semiconductor based power supplies, phase controllers, reactors, etc are used enormously in modern power system network and they are contributing huge amount of harmonics to the power system. In electric power system, main sources of harmonics may be classified as follows [4–11]

1. Magnetization nonlinearities of transformer
2. Rotating machines
3. Arcing devices
4. Semiconductor based power supply system
5. Inverter fed A.C. drives
6. Thyristor controlled reactors
7. Phase controllers
8. A.C. regulators

Above mentioned sources are described in the following sections.

4.4.1 Magnetization Nonlinearities of Transformers

Transformer magnetic material characteristic is non linear. This non linearity is the main reason for harmonics during excitation. Sources of harmonics in transformer may be classified into four categories as follows:

1. **Normal Excitation:** Normal excitation current of a transformer is non sinusoidal. The distortion is mainly caused by zero sequence triplen harmonics and particularly the third present in the excitation current. Presence of the electric path like air, oil or tank for zero sequence components can be used to reduce those harmonics. Their high reluctance tends to reduce them. Delta connection of poly-phase transformer is very effective to reduce triplen harmonics provided the three phase voltages are balanced.
2. **Symmetrical Over Excitation:** Transformers are designed to make good use of the magnetic properties of the core material. When such transformers are subjected to a rise in voltage, the cores face a considerable rise in magnetic flux density, which often causes considerable saturation. This saturation with symmetrical magnetizing current generates all the odd harmonics. The fundamental component is not a problem and all triplen harmonics can be absorbed by delta connection in balanced system. The harmonics generated by symmetrical over excitation are odd harmonics (like 5, 7, 11, 13, 17, 19.... etc) i.e. those of orders $6k \pm 1$, where k is an integer.
3. **Inrush Current Harmonics:** When a transformer is switched off, sometimes there exists a residual flux density in the core. When the transformer is re-energized the flux density can reach peak levels of twice the maximum flux density or more. It produces high ampere-turns in the core. This causes magnetizing currents to reach up to 5–10 per unit of the rated value, which is very high as compared to the normal values of a few percentage points. This is known as inrush current. This causes generation of enormous second order harmonic component in the transformer current.
4. **D.C. Magnetization:** Under magnetic imbalance, the shape of the magnetizing characteristics and the excitation currents are different from those under no load conditions. When the flux is unbalanced, the core contains an average value of

flux (ϕ_{dc}), which is equivalent to a direct component of excitation current of the transformer. Under such unbalance conditions, the transformer excitation current contains both odd and even harmonic components.

4.4.2 Rotating Machine

Rotating machines also act as source of harmonics in power system. Causes of harmonics generation in rotating electrical machines are classified into following categories:

1. **Magnetic nonlinearities of the core material:** nonlinear magnetization characteristics of the core material causes harmonic generation.
2. **Non uniform flux distribution in air gap:** often it is assumed that the air-gap flux distribution is uniform and the operating principles of rotating machines are discussed based on this assumption. But in most of the rotating machines, flux distribution in air-gap is not uniform which leads to harmonics production.
3. **Slot harmonics:** slots are inevitable in rotating machines. Alternate presence of slot and teeth changes the reluctance of the magnetic flux varies in similar type of alternating fashion. This variation acts as a reason for harmonic generation. Harmonics produced due to pitch factor and distribution factor.
4. **Design parameters like pitch factor and distribution factors:** harmonic generation in synchronous generator depends on the different design factors like pitch factor and distribution factors.
5. **Rotor saliency:** rotor saliency brings the variation of reluctance in the magnetic path and reactance in electric path which contribute to the harmonic generation.
6. **Crawling:** it is a common problem faced by induction motors. During this fault, odd harmonics like 5th and 7th orders appear. Fifth harmonics rotates in the same direction as of the fundamental but 7th order harmonics rotate in opposite to this direction. It changes the operating characteristics of the motor. During crawling, 7th harmonics is dominating over the 5th harmonics and lowers the stable operating zone to one seventh speed of the rated value.
7. **Cogging:** It is a problem where induction motors fail to start at all. Harmonic production from the motor during this condition is different from the normal condition.
8. **Rotor misalignment:** rotor misalignment causes variation of flux linkage in each cycle of rotation contributing to harmonic generation.
9. **Mass unbalance:** with the aging, mass unbalance is observed specially in the rotor side. This refers to the core property and adds in harmonic generation.
10. **Bar breakage:** bar breakage in an induction motor, either symmetrically or asymmetrically, is reflected in harmonic generation in rotor circuit as well as in stator side.
11. **Fractal error:** fractal error in core material produces harmonics.
12. **Unsymmetrical fault:** unsymmetrical fault is also a reason for harmonic generation related to negative sequence components.

4.4.3 Distortion Caused by Arcing Devices

Arcing devices are very important source of power system harmonics. The voltage versus current characteristics of an electric arc in an arcing device are highly non linear. Arc ignition is equivalent to a short circuit current with decrease in voltage. The voltage-current is controlled by the power system impedance. In respect of harmonic generation, arcing devices are divided into three main categories:

1. electric arc furnace
2. discharge type lighting
3. arc welders.

4.4.4 Power Supplies with Semiconductor Devices

Semiconductor based power supply systems are the main sources of harmonics. Harmonics generated in power supply include integer harmonics, inter harmonics and sub harmonics. Frequencies and magnitudes of the harmonics depend on the type of semiconductor devices used in the power supplies, operating point, nature of load variation, etc.

4.4.5 Inverter Fed AC drives

Application of AC drives has increased to a great extent, most of which are inverter fed AC drives. They use switching circuits using semiconductor devices like GTO, IGBT, etc. Pulse width modulation (PWM) has got very popularity in AC drive application. All these drives are sources of integer as well as fractional harmonics.

4.4.6 Thyristor Controlled Reactors

VAR compensators used in power system network are also source of harmonics. Different types of thyristor controlled reactors are used in power system like series controller, shunt controller, static VAR compensator (SVC), fixed capacitor thyristor controlled reactor (FCTCR), thyristor switched capacitor thyristor controlled reactor (TSCTCR). All these circuits are sources of harmonics in power system. Use of static synchronous generator (SSG), voltage source STATCOM, current source STATCOM, etc in power system are increasing rapidly. All these contribute harmonics of both integer and fractional type in power system. For example, SVC produces odd harmonics. Under perfectly symmetrical voltage conditions, triplen harmonics are kept out of the line by delta connection.

4.4.7 Phase Controller

For supply of stable and balanced three phase electric power, phase controller plays important role in power system. Phase controllers used in power system act as source of harmonics. Modulated phase control method is used in cyclo-converter. It performs static power conversion from one frequency to another frequency. Most of the cyclo-converter waveforms contain frequencies which are not integer multiples of the main output frequency.

4.4.8 AC Regulators

AC regulators used in power system apply both off line and on line control technique for voltage regulation which result in harmonic generation. On line regulation technique distorts wave-shape more than off line regulation along with other power system disturbances like transients, DC offset, flicker etc. Thyristor controlled single phase or polyphase regulators using half wave, full wave or integral cycle control technique produce sub-harmonics and inter-harmonics in power system.

4.5 Effects of Harmonics

In electrical power system, harmonics are not desirable in most of the applications and operations. Harmonics have adverse effect on power system equipment as well as on its operation. Harmonics can create resonance in power system network. Damping property may change due to the presence of harmonics. Also it has some adverse effects on performance of rotating machines, transformers and transmission networks. Accuracy and operating characteristics of measuring instruments and protective devices may change due to the presence of undesirable harmonics. Performance of reactive power compensation devices may change. Moreover presence of harmonics has some adverse effects on different consumer equipment. Effects of harmonics are classified in the following way [7–12]:

1. Resonance
2. Poor Damping
3. Effects of Harmonics on Rotating Machines
4. Effects of Harmonics on Transformer
5. Effects of Harmonics on Transmission Lines
6. Effects of Harmonics on Measuring Instruments
7. Harmonic Interference with Power System Protection
8. Effects of Harmonics on Capacitor Banks
9. Effects of Harmonics on Consumer Equipment

4.5.1 Resonance

Capacitors used for power factor correction cause system resonances due to harmonic frequencies. This results in excessive high current, which can produce damage to the capacitors. Resonance occurs when the frequency at which the capacitive and inductive reactance of the circuit impedance are equal. At the resonant frequency, a parallel resonance has high impedance and series resonance low impedance. Harmonic resonances create problems in operation of power factor correction capacitors.

There are three types of resonance occurred in power system

1. **Parallel Resonance,** which offers high impedance at the resonance frequency and increases harmonic voltages and harmonic currents.
2. **Series Resonance,** which results in high capacitor current at relatively small harmonic voltages. Magnitude of this current depends upon the quality factor of the circuit.
3. **Complementary and Composite Resonance:** Normally, composite resonant frequency is of non-integer type frequencies generated from conversion from the fundamental frequency and d.c. components of the converter control circuit.

4.5.2 Poor Damping

In presence of harmonics, variable speed drive motors or a switched mode power supply introduces small negative impedance or resistance. It decreases the current with the rise of voltage, which reduces the damping or broadband energy absorption capability of the system. Undesirable variation of the degree of damping changes the operating performance of different electrical devices, especially different measuring and controlling instruments.

4.5.3 Effects of Harmonics on Rotating Machines

1. **Harmonic losses:** Harmonic voltages or currents increase losses in the stator windings, rotor circuit, and stator and rotor lamination. Normally the losses in the stator and rotor conductors of A.C. machines are greater than those associated with the D.C. resistances due to the presence of eddy current and skin effect in ac circuit. Harmonics increase both copper loss and core loss. This results in overheating and reduction of the efficiency of a machine.
2. **Harmonic torque:** Harmonic currents present in the stator of an a.c. machine produce induction motoring action (i.e. positive harmonic slips S_n), which gives rise to shaft torques in the same direction as the harmonic field velocities in such a way that all positive sequence harmonic will develop shaft torques aiding shaft rotation whereas negative sequence harmonics will have the opposite effect.
3. **Speed torque characteristics:** Each harmonic component contributes to the magnetic force. The presence of harmonics has effect on speed/torque characteristic.

4. **Cogging:** Cogging is the failure of an induction motor to run up to normal speed due to a stable operating point occurring at a lower frequency.
5. **High capacitive current:** The presence of harmonics increases capacitive current through the stray capacitance in ASD-fed electric motors which is one of the reasons for their failure.
6. **Voltage stress on insulation:** Harmonic voltages increase the stress on insulating materials.

4.5.4 Effects of Harmonics on Transformers

Harmonics has effect on transformer in various ways, e.g.:

1. **Core loss:** Harmonic voltage increases the hysteresis and eddy current losses in the laminations. The amount of the core loss depends on harmonic present in supply voltage design parameter of core materials and magnetic circuit.
2. **Copper loss:** Harmonic current increases copper loss. The loss $\left(\sum_{n=2}^{\alpha} I_n^2 R_n\right)$, mainly depend on the harmonics present in the load and effective ac resistance of the winding. Copper loss increase temperature and create hot spots in that transformer. The effect is prominent in the case of converter transformers these transformers do not benefit from the presence of filters as filter are normally connected on the a.c. system side.
3. **Stress:** Voltage harmonics increase stresses of the insulation,
4. **Core vibration:** Current and voltage harmonics increase small core vibrations.
5. **Saturation problem:** Sometimes additional harmonic voltage causes core saturation.

Guidelines for transformer derating with respect to the harmonic current are given in the ANSI/IEEE standard C57.110.

4.5.5 Effects of Harmonics on Transmission System

Harmonics has effect on transmission line in various ways.

1. **Skin effect and Proximity effect:** these effects depend on frequency. Harmonics increase these effects. As a result effective ac resistance increases in presence of harmonics.
2. **Loss:** Additional harmonic current increases copper loss of transmission system and reduce power transmitting capacity. Copper loss is given by $\sum_{n=2}^{\alpha} I_n^2 R_n$, where I_n is the nth harmonic current and R_n is the system resistance at that harmonic frequency.
3. **Voltage drop:** Harmonic current produces harmonic voltage drops across various circuit impedances. As a result a weak system of large impedance has low fault level and greater voltage disturbances where as a stiff system of low impedance has high fault level and lower voltage disturbance.

4. **Dielectric stress:** Harmonic voltage increases dielectric stress of cables used in transmission line. The dielectric stress is proportional to their crest voltages. This reduces dielectric strength. This results in shortening of the useful life of the cable and probability of the number of faults and hence, the cost of repairs.
5. **Corona:** Corona starting and extinction levels depend on peak to peak voltage. The peak voltage depends on the phase relationship between the harmonics and the fundamental. Harmonic voltage increases peak to peak voltage. It may happen that the peak voltage is above the rating while the r.m.s. voltage is well within this limit. In this matter the IEEE 519 standard provides typical capacity derating curves for cables feeding six pulse converters.

4.5.6 Effects of Harmonics on Measuring Instruments

Harmonics has effect on measuring instruments in various ways

1. **Error:** Measuring instruments are calibrated on purely sinusoidal alternating current but they are used on a distorted electricity supply. This introduces error in measurement.
2. **Sign of error:** Sign of error depends on the magnitude and the direction of the harmonic power.
3. **Harmonic torque:** Torque produced by harmonics greatly affects operation of instruments.
4. Any d.c. power supplied to or generated by the customer will cause an error proportional to the harmonic-fundamental power ratio, with the error sign related to the direction of power flow.
5. Harmonic voltages or currents not only produce torques, but also degrade the capability of a meter to measure fundamental frequency power.
6. The kilowatt-hour meter, based on the Ferraris (eddy current) motor principle, show generally appreciably high readings with a consumer generating harmonics through thyristor-controlled variable speed equipment particularly where even harmonics and d.c. are involved. By this way, consumers that generate harmonics are automatically penalized by higher apparent electricity consumption. This may well offset the supply authority's additional losses. It is therefore in the consumer's own interest to reduce harmonic generation to the greatest possible extent.
7. There is no proof or evidence that the reading of kVA demand meters is affected by network harmonics.
8. KW demand meters operating on the time interval Ferraris motor principle show high reading in presence of harmonics.
9. Harmonics create problems in measurement of VAR values in power networks as VAR is a quantity defined with respect to sinusoidal waveforms.
10. Absolute average and peak responding meters which are calibrated in r.m.s. are not suitable in the presence of harmonic distortion.

4.5.7 Harmonic Interference with Power System Protection

1. Harmonics degrade the operating characteristics of protective relays. Effect of harmonics on relay depends on the design features and principles of operation.
2. Some digital relays and algorithms operate on sample data and zero crossing moment. Harmonic distortion creates error on such operation.
3. Harmonics make higher di/dt at zero crossings and the current sensing ability of the thermal magnetic breakers and change trip point due to extra heating in the solenoid.
4. Current harmonic distortion affects the interruption capability of circuit breakers and fuses.
5. The fuses are thermally activated and inherently r. m. s. over current devices. Fuse materials are also susceptible to extra skin effect of the harmonic frequencies.

However, the changes in operating characteristics are small and do not present a problem. Most studies say that it is difficult to predict relay performance without testing. A lot of studies have been published on electro-mechanical and electronic relays. Presence of harmonic current, particularly third harmonic, in a fault situation, results in considerable measurement errors relative to the fundamental based setting. Harmonic content increases the possibility of mal-operation, unless only the fundamental waveforms are captured.

4.5.8 Effects of Harmonics on Capacitor Banks

1. **Dielectric loss:** Harmonic voltage increases the dielectric loss in capacitors.
2. **Thermal stress:** Harmonic voltage increases the stress on capacitor bank and the system connected to the bank.
3. **Resonance:** Harmonics cause series and parallel resonances between the capacitors and the rest of the system. This results in over voltage and high currents, which increase the losses and overheating of capacitors, and often leading to their destruction.
4. **Reactive power:** Change of harmonic contents sometimes increases reactive power over permissible manufacturer tolerances.
5. **Power factor correction capacitors:** Presence of harmonics causes malfunction of the operation of power factor correction capacitors.

4.5.9 Effects of Harmonics on Consumer Equipment

IEEE Task Force on the Effects of Harmonics on Equipment has made a wide study on this matter. The result can be summarized as follows:

1. Television Receivers: Harmonics changes in TV picture size and brightness. Inter-harmonics changes amplitude modulation of the fundamental frequency. For example, even a 0.5% inter-harmonic level can produce periodic enlargement and reduction of the image of the cathode ray tube.
2. Fluorescent and mercury arc lighting: capacitors used in such lighting applications together with the inductance of the ballast and circuit produce a resonant frequency. It results in excessive heating and failure in operation. Audible noise is produced due to harmonic voltage distortion.
3. Computers: harmonics create problems in monitor and CPU operation. Harmonic rate (geometric) measured in vacuum must be less than −3% (Honeywell, DEC) or 5% (IBM). CDC specifies that the ratio of peak to effective value of the supply voltage must equal to 1.41 ± 0.1.

4.5.10 Summary of Effects of Harmonics

Effects of harmonics in power system discussed in the previous sections can as summarized as given in Table 4.2.

Table 4.2 Effects of harmonics on different electrical components

Name of component	Effects of harmonics
Generator	Production of pulsating or oscillating torques which involve torsional oscillations of rotor elements of TG set and rotor heating
Motor	Stator and rotor copper losses increase due to harmonic current flow, leakage flux created by harmonic currents causes additional stator and rotor losses, core loss increases due to harmonic voltages and positive sequence harmonics develop shaft torques that aid shaft rotations whereas negative sequence opposes it
Transformer	Stray losses increase due tom harmonic current flow, hysteresis losses increase, due to presence of high frequency harmonics resonance may occur between winding inductance and line capacitance
Relaying	Mal-tripping may occur due to presence of harmonics which affects the time delay characteristics
Switchgear	Due to predominance of skin and proximity effects at higher frequencies, bus-bars behave like cables and transient recovery voltage changes which affect the operation of blow-out coils
Capacitor	Due to presence of harmonics, reactive power increases, dielectric losses increase causing additional heating and resonance and overvoltage may occur, resulting in reduced life
Cables	Due to increased skin and proximity effects at higher frequencies, additional heating occurs, R_{ac} increases and ac copper loss increases
Consumer equipment	Life and efficiency reduce drastically
Communication circuits	Noise creeps in transmitted signals

4.6 Harmonic Standard

Different standards have been prepared from time to time by IEEE, ANSI, CBEMA which provide guidelines for power quality usages and practices. Harmonic standards mainly include the following issues:

1. description and characterization of the phenomenon
2. major sources of harmonic problems
3. impact on other equipment and on the power system
4. indices and statistical analysis to provide a quantitative assessment and its significance
5. measurement techniques and guidelines
6. emission limits of quality degradation for different types and classes of equipment
7. immunity or tolerance level of different types of equipment
8. testing methods and procedures for compliance with limits'
9. mitigation guidelines

4.6.1 The IEC Standard

Geneva based International Electro-technical Commission or Commission Electrotechnique Internationale (IEC) is the widely recognized organization as the curator of electric power quality standards. IEC has introduced a series of standards, known as Electro-Magnetic Compatibility (EMC) standards, to deal with power quality issues. Integer and inter harmonics are included in IEC61000 series as one of the conducted low-frequency electro-magnetic phenomena. The series also provides internationally accepted information for the control of power system harmonic (and inter-harmonic) distortion [7].

IEC 61000 1-4: It provides the rationale for limiting power frequency conducted harmonic and inter-harmonic current emissions from equipment in the frequency range up to 9 kHz.

IEC 61000 2-1: Power system equipment, industrial loads and residual loads as considered here as three major sources of harmonics in this series. The series consider HVDC converters, FACTS devices as main power system equipment for harmonic distortion originating in the transmission system. Static power converters and electric arc furnaces are includes as harmonics sources in the industrial category. Appliances powered by rectifiers with smoothing capacitors like PCs, TV receivers, etc are considered as the main distorting components in the residential category.

IEC 61000 2-2: Compatibility levels of the harmonic and inter-harmonic voltage distortion in public low-voltage power industry systems are discussed in this section.

IEC 61000 2-4: Compatibility levels of harmonic and inter-harmonic for industrial plant with the main effects of inter-harmonics are discussed in this section.

IEC 61000 2-12: This section covers compatibility levels for low-frequency conducted disturbances related to medium voltage power supply systems and injected signals such as those used in ripple control equipment.

IEC 61000 3-2 and 3-4: this section gives the guideline for limiting harmonic current emissions by equipment with input currents of 16 A and below per phase along with measurement circuit, supply source and testing conditions as well as the requirements for the instrumentation for harmonics monitoring.

IEC 61000 3-6: This section deals with the capability levels for harmonic voltages in low and medium voltage networks as well as planning levels for MV, HV and EHV power systems indicating emission limits for distorting loads in MV and HV power systems.

IEC 61000 3-12: Limits for the harmonic currents produced by equipment connected to low voltage systems with input currents equal to and below 75 A per phase are discussed in this section.

IEC 61000 4-7: This section deals with testing and measurement techniques.

IEC 61000 4-13: Testing and measurement techniques with reference to harmonics and inter-harmonics, including mains signaling at a.c. power ports as well as low-frequency immunity tests are covered in this section.

4.6.2 IEEE 519-1992

IEEE 519-1992 document is a widespread alternative to the IEC series. It identifies the major sources of harmonic in power systems. Following main features are referred in this series [7]:

- Power converters, arc furnaces, static VAR compensators, inverters of dispersed generation, electric phase control of power, cyclo-converters, switch mode power supplies and pulse width modulated (PWM) drives are considered as major sources for harmonics in power system.
- Wave shape distortion due to the harmonics is explained.
- Response in presence of harmonics and different resonance occurred by harmonics are covered.
- Distortion caused by harmonics in low-voltage distribution system, industrial systems and transmission systems are discussed.
- Effects of harmonic distortion on the operation of various devices or loads which comprise motors and generators, transformers, power cables, capacitors, electronic equipment, metering equipment, switchgear, relays and static power converters are covered.
- Interference to the telephone networks is covered.
- Recommendation for reducing the amount of telephone interference caused by harmonic distortion in the power system is done.
- Guidelines for measurement of harmonic distortion in the power system are given.
- Useful calculation of harmonic currents, system frequency responses and modeling of various power system components for the analysis of harmonic propagation are mentioned.
- Currently available harmonic monitoring techniques is referred.

Table 4.3 Different power quality parameters and their definitions

EPQ parameter	Definition
Total Voltage Harmonic Distortion (THD_V)	$\left(\sqrt{\sum_{n=2}^{n=\infty} V_n}\right) / V_1$
Total Current Harmonic Distortion (THD_I)	$\left(\sqrt{\sum_{n=2}^{n=\infty} I_n}\right) / I_1$
Active power distortion factor	$PDF = \frac{P_H}{P_1} \times 100\%$
Reactive power distortion factor	$QDF = \frac{Q_H}{Q_1} \times 100\%$
Apparent power distortion factor	$ADF = \frac{A_H}{A_1} \times 100\%$
Flicker factor	$\Delta V / \lvert V \rvert$
Crest factor	V_{peak} / V_{RMS}
PQ index	$\left[\left\lvert \frac{\text{V} - 100\%}{V_{CBEMA}(t) - 100\%} \right\rvert\right] .100\%$

- Techniques for smoothing-out the rapidly fluctuating harmonic components are given.
- Different design aspects including reactive power compensation for systems with harmonic distortion, are guided.
- Various techniques for reducing the amount of harmonic current penetrating into the a.c. systems are given.
- Recommendation and useful suggestion for both individual consumers and utilities for controlling the harmonic distortion to tolerable levels are made in this section.

4.6.3 General Harmonic Indices

Different useful definitions can be summarized as given in Table 4.3 [2, 13–19].

References

[1] De Rosa, F., Langella, R., Sollazzo, A., Testa, A.: On the inter-harmonic components generated by adjustable speed drives. In Proceedings 10th IEEE International Conference Harmonics Quality Power, Rio de Janeiro, 6–9 Oct 2002

[2] Li A., Xu, W.: On defining harmonic contributions at the point of common coupling. IEEE Power Eng. Rev. **22**(7), 44–45 (2002)

[3] Xu, W.: Component modeling issues for power quality assessment (Invited Paper). IEEE Power Eng. Rev. **21**(11), 12–17 (2001)
[4] Carbone, R., Rosa, F.D., Langella, R., Testa, A.: A new approach for the computation of harmonics produced by line-commuted AC/DC/AC converters. IEEE Trans. Power Deliver. **20**(3), 2227–2234 (2005)
[5] Shmilovitz, D.: On the definition of total harmonic distortion and its effects on measurement interpretation. IEEE Trans. Power Deliver. **20**(1), 526–528 (2005)
[6] Li, C., Xu, W., Tayjasanant, T.: A critical impedance based method for identifying harmonic sources. IEEE Trans. Power Deliver. **19**(2), 671–678 (2004)
[7] Arrillaga, J., Watson N.R.: Power System Harmonics. Wiley, Chichester (2003)
[8] Domijan, A., Heudt, G.T., Meliopoulos, A.P.S., Venkata, S.S., West, S.: Direction of research on power quality. IEEE Trans. Power Deliver. **8**(1), 429–436 (1993)
[9] Dorr D.S.: Power Quality Study—1990–1995 Initial Results. Proc. IEEE, **APEC-92**, 303–308 (1992)
[10] Emanuel, E., Orr, J.A., Galachenski, E.M.: A survey of harmonic voltage and currents at distribution substations. IEEE Trans. Power Deliver. **5**(4), 1883–1890 (1991)
[11] Subjak J.S., Jr., Mcquilkin, J.S.: Harmonics-causes, effects, measurements, analysis: An update. IEEE Trans. Ind. Appl. **26**, 1034–1042 (1990)
[12] Emanuel, A.E., Orr, J.A., Cyganski, D., Gulchenski, E.M.: A survey of harmonics voltages, currents at the customer's bus. IEEE Trans. Power Deliver. **8**, 411–421 (1993)
[13] Jaramillo, S.H., Heydt, G.T., O'Neill-Carrillo, E.: Power quality indices for aperidic voltages and currents. IEEE Trans. Power Deliver. **15**(2), 784–790 (2000)
[14] Xu, W., Liu, Y.: A method for determining customer and utility harmonic contributions at the PCC. IEEE Trans. Power Deliver. **15**(2), 804–811 (2000)
[15] IEEE Std: Definitions for the Measurement of Electric Quantities under sinusoidal, non sinusoidal, balanced or unbalanced conditions. IEEE Std. 1459–2000, Jan. 2000
[16] Sabin, D.D., Brooks, D.L., Sundaram, A.: Indices for assessing harmonics distortion from power quality measurement: Definitions and benchmark data. IEEE Trans. Power Deliver. **14**(2), 489–496 (1999)
[17] Carbone, R., Menniti, D., Morison, R.E., Testa, A.: Harmonics and inerharmonic distortion in current source type inverter drives. IEEE Trans. Power Deliver. **10**(3), 1576–1583 (1995)
[18] IEEE Standard: IEEE Recommended Practice for Monitoring Electric Power Quality. IEEE Standard 1159–1995 (1995)
[19] Hu, L., Yacamini, R.: Calculation of harmonics and interharmonics in HVDC schemes with low DC side impedance. Proc. Inst. Elect. Eng. C. **140**(6), 469–475 (1993)

Chapter 5
Transients

Abstract This chapter presents transients as a power quality problem. It then describes the events behind the transients. Classification of power system transients is done on the basis of nature and events. Effects of transients are mentioned in short.

5.1 Introduction

Transients are short duration power system phenomena. They show very high rise of voltage and current in short period of time. Power system transients may generate in the system itself or may propagate to the system from other external system. Transients are treated as one major disturbance related to power quality aspects [1]. They may be unidirectional or oscillatory in shape. Though such events are of short duration they create very high magnitude of voltage and/or current. Transients affect the performance of power system devices as well as measuring, controlling and protective devices. Recently transients are viewed as a potential power quality problem. This increases the requirements on characterization and analysis of transient waveforms. Deep study is needed to establish the relation between waveform characteristics and equipment performance and to find out the causes and effects of transients.

5.2 Power System Transients

The term "transient" denotes the voltage and current components that occur during the transition from one steady state (typically sinusoidal) to another steady state. Homogenous solutions of differential equations describing electrical circuits are sum of a homogenous solution which correspond with the transient. In such solutions, a transient is always associated with a change in steady state due to a switching action.

However, most of the power system transients refer to short duration events of voltage and current [2–4]. There is no clear limit for the period of transient. Generally,

S. Chattopadhyay et al., *Electric Power Quality*, Power Systems,
DOI 10.1007/978-94-007-0635-4_5,

short duration phenomena of less than one cycle (referred to as power system frequency, 50 or 60 Hz) are generally referred to as transients. Accurate modeling of the power system transients and the characterization of measured transients along with their sources and effects are very important [5–7].

5.3 Causes of Power System Transients

There are different causes for power system transients. For example, lightning strokes to the wires in the power system or to ground and component switching either of network components or end user equipment can produce transients. Nature of power system transients are very much event dependent [8–10].

Short duration events can be classified into three classes:

1. Events that can be identified by their fundamental frequency magnitude. Voltage magnitude in such events goes through significant changes for long periods. The changes are well apart and observable with respect to time. This enables magnitude estimators to identify and resolving events having significant changes. These are observed mainly in fault induced events, transformer saturation, induction motor starting, etc. Voltage dips with duration typically between 50 ms and several seconds and interruptions with duration from several seconds up to many hours are associated with such transient events.
2. Events having significant changes in the fundamental frequency magnitude but of short duration. In such events, it is very difficult to extract voltage magnitude of transients. These are normally observed in fuse-cleared faults and self-extinguishing faults.
3. Events of very short duration (transients) for which the fundamental frequency magnitude does not offer important information. For this class, the higher frequency components of the signal must be considered for a thorough characterization and classification.

Based on waveform shape, power system transients, can be classified into

1. Oscillatory transients
2. Impulsive transients
3. Multiple transients

Transients with their probable causes are summarized in Table 5.1.

Table 5.1 Classification of transients

Waveform based classification	Event based classification
Impulsive transients	Lightning
Oscillatory transients	Capacitor energizing
	Re-strike during capacitor de-energizing
	Line or cable energizing
Multiple transients	Current chopping
	Multiple re-strikes
	Repetitive switching actions

5.3.1 *Impulsive Transients*

An impulsive transient is defined as a sudden change in the steady state condition of voltage, current or both, that is unidirectional in polarity either positive or negative. Analysis of impulsive transients is done by their rise and decay times. Impulsive transients are damped quickly by the resistive circuit elements and do not propagate far from their source. Thus their effects are localized.

Impulsive transients are common during lightning. Lightning stroke may appear directly or by indirect induction. When a lightning stroke hits a transmission line (direct stroke) an impulsive over voltage is induced. They have high magnitude. Lightning overvoltage can also be induced by nearby strokes to the ground or between clouds. These over-voltages are of lower magnitude than those produced by direct strokes. Normally impulsive transient shows a sudden rise followed by an exponential decay. But in some cases, lightning transient shows a sudden rise followed by a sudden drop and an oscillation with relatively small amplitude.

5.3.2 *Oscillatory Transients*

Oscillatory transient is alternating in nature. It shows a damped oscillation with a frequency ranging from a few hundred Hertzs up to several Megahertzs. Oscillatory transients can be mathematically derived by the homogenous solution to linear differential equations. As the electric power system can approximately be described by a set of linear differential equations, oscillatory transients are the "natural transients" in electric power system. For this reason, oscillatory transients dominate over impulsive transients. For example, oscillatory transient can be caused by the energizing of a capacitor bank where, frequency of oscillation is mainly determined by the capacitance of the capacitor bank and the short circuit inductance of the circuit feeding the capacitor bank (capacitor energizing). Another common cause of oscillatory transient is event of energizing of transmission line.

5.3.3 *Multiple Transients with a Single Cause*

Multiple transients are combination of many overlapped transients occurred due to more than one switching action. For example, in three phase system the switching action in the individual phases rarely take place at the same instant. Such events produce multiple transients. Current chopping and restrike are other two major causes of multiple transients. Current chopping is done when the current during opening of a circuit breaker becomes zero before the natural zero crossing. This results in transients of high over-voltages. Restrike may occur when a capacitor is de-energized by a slowly moving switch where the voltage over the capacitor increases faster than the voltage-withstand of the gap between the contacts of the switch.

5.4 Effects

Transients are very much related to the operation and performance of different parts of power system as well as loads and measuring and protective devices also. Nature and duration of /power system transients are related to correct operation of circuit breakers, and overvoltage due to switching of high voltage lines. High magnitudes of voltage transients break insulations of the system. High magnitude of current transients can burn out devices and instruments. Transients can cause mal-operation of relays and mal-tripping of circuit breakers. Frequent number of direct or indirectly induced oscillatory transients may change the magnetic properties of core materials used in electric machines.

References

[1] Arrillaga, J., Watson, N.R., Chen, S.: Power System Quality Assessment. New York, Wiley (2000)
[2] Herath, C., Gosbell, V., Perera, S.: A transient index for reporting power quality (PQ) surveys. In Proceedings of CIRED 2003, Bercelona, pp. 2.61-1–2.61-5 (2003)
[3] Huang, S.-J., Yang, T.-M., Huang, J.-T.: FPGA realization of wavelet transform for detection of electric power system disturbances. IEEE Trans. Power Deliv. **17**(2), 388–394 (2002)
[4] Shaw, S.R., Laughman, C.R., Leeb, S.B., Lepard, R.F.: A power quality prediction system. IEEE Trans. Ind. Electron. **47**(3), 511–517 (2000)
[5] Santoso, S., Lamoree, J., Grady, W.M., Powers, E.J., Bhatt, S.C.: A scalable PQ event identification system. IEEE Trans. Power Deliv. **15**, 738–742 (2000)
[6] Santoso, S., Powers, E.J., Grady, W.M., Parsons, A.C.: Characterization of distribution power quality events with fourier and wavelet transform. IEEE Trans. Power Deliv. **15**(1), 247–256 (2000)
[7] Bollen, M.H.J.: Understanding power quality problems. IEEE Press Series on Power Engineering, New York (2000)
[8] Martzloiff, F.D., Gruzs, T.M.: Power quality site surveys: Facts, fiction, and fallacies. IEEE Trans. Ind. Appl. **24**(6), 1005–1018 (1998)
[9] Simpson, R.H.: Instrumentation, measurement techniques and analytical tools in power quality studies. Proceedings of IEEE Annual Conference of Pulp and Paeu Industry, pp. 119–140 (1997)
[10] Long, W., Cotcher, D., Ruiu, D., Adam, P., Lee, S., Adapa, R.: EMTP—a powerful tool for analyzing power system transients. IEEE Comput. Appl. Power. **3**(3), 36–41 (1990)

Chapter 6
Sag, Swell, Interruption, Undervoltage and Overvoltage

Abstract This chapter highlights the power system disturbances related to rise or fall of rms voltage. Then it describes these disturbances one after another along with their main causes and effects.

6.1 Introduction

Other than harmonics, transients and unbalance, power system suffers from the disturbances of change of rated voltage (or current). These disturbances include rise and fall of voltage for certain period of time. Based on the rise or fall of rms voltage in terms of per unit along with its duration, they are classified as sag or voltage dips, swell, interruption, under voltage and over voltage. These disturbances are very common and have some measure with respect to power quality issues [1–3]. However, monitoring of such disturbances is not too difficult because it depends on the monitoring of rms values in time frame.

6.2 Sag

Sag is reduction of rms voltage for short duration to an extent between 0.1 and 0.9 pu [4, 5]. Duration of sag is considered in between 0.5 cycles and 1 min. Sag is often called as voltage dip.

There are varieties of causes for voltage sags, such as

- **Energization of heavy loads:** sudden energization of heavy load reduces voltage. If the supply is capable of delivering this high load, then bus voltage level quickly gets back to its rated value. Example of such high load is arc furnace. Connection of arc furnace may cause sag or voltage dip in power system.
- **Starting of large induction motors:** polyphase induction motors draw high current at starting. Thus connection of large poly phase induction motors to a bus often causes sag or voltage dips in power system because of high starting current.
- **Single line-to-ground faults:** high fault current because of single line to ground (SLG) fault reduces bus voltage suddenly causing sag or voltage dip in power system.

S. Chattopadhyay et al., *Electric Power Quality,* Power Systems,
DOI 10.1007/978-94-007-0635-4_6,

- **Line-line and symmetrical fault:** this fault also reduces voltage causing sag in power system.
- **Load transferring from one power source to another:** at the time of load transferring from one source to another or from one phase to another, voltage dip or sag may occur in the power system.

Effects of sag mainly includes:

- Voltage stability because of reduction of bus voltage for short duration
- Malfunctions of electrical low-voltage devices
- Malfunctions of uninterruptible power supply
- Malfunction of measuring and control equipment
- Interfacing with communication signals

6.3 Swell

Swell is opposite of sag. It is a short duration phenomenon of increase in rms voltage. Voltage magnitude lies between 1.1 and 1.8 pu and duration of the event ranges from 0.5 cycles to 1 min. Swells are rare events as compared to sags.

Main causes of swell are:

- **Switching off of a large load:** sudden reduction of large loads by switching off causes swell in the power system.
- **Energizing a capacitor bank:** capacitor bank draws leading current. Voltage increases during enegization of capacitor bank which may cause swell.
- **Voltage increase of the unfaulted phases during a single line-to-ground fault:** in single line to ground fault in an ungrounded power system, voltages of healthy phases increase which may cause swell in those phases.
- **"Momentary overvoltage":** it is often used as a synonym for the term swell. In fact momentary overvoltage due to power frequency surge or transients may cause swell.

Like sag, effects of swell mainly includes

- Voltage stability because of reduction of bus voltage for short duration
- Malfunctions of electrical low-voltage devices
- Malfunctions of uninterruptible power supply
- Malfunction of measuring and control equipment
- Interfacing with communication signals

6.4 Interruption

Interruption is an event of reduction of supply voltage (or load current) to less than 0.1 pu.

Causes:

There are many reasons for power interruption. Some of the general causes of interruption are:

- equipment failures
- control malfunction
- blown fuse
- breaker opening

Depending on the duration of reduction of voltage, interruption is classified into many groups,

According to the European standard EN-50160:

1. A short interruption is up to 3 min;
2. A long interruption is longer than 3 min.

Based on the standard IEEE-1250:

1. An instantaneous interruption is between 0.5 and 30 cycles;
2. A momentary interruption is between 30 cycles and 2 s;
3. A temporary interruption is between 2 s and 2 min; and
4. A sustained interruption is longer than 2 min.

6.5 Sustained Interruption

Among all interruption, sustained (or long) interruption is most severe. In such interruption voltage drops to zero and it does not return automatically. According to the IEC definition, the duration of sustained interruption is more than 3 min; but based on the IEEE definition the duration is more than 2 min. Frequency of occurrence and duration of long interruptions are the two main aspects in assessing the ability of a power system to deliver reliable service to customers.

The most important causes of sustained interruptions are:

- Fault occurrence in a part of power systems
- Incorrect intervention of a protective relay
- Scheduled (or planned) interruption in a low voltage network with no redundancy.

6.6 Undervoltage

The undervoltage is reduction of rms voltage to 0.8–0.9 pu for more than 1 min. Normal duration of undervoltage is greater than sag. Main causes of undervoltage are overload, less supply capability, fault, etc. Undervoltage may cause voltage instability, drawl of high current by motors, high reactive power demand, etc.

6.7 Overvoltage

The over voltage is increase of rms voltage to 1.1–1.2 pu for more than 1 min. Normal duration of undervoltage is greater than swell. There are many reasons for occurring overvoltage in power system as follows:

- Overvoltages generated by an insulation fault
- Ferroresonance
- Faults with the alternator regulator, tap changer transformer, or overcompensation
- Lightning overvoltages
- Switching overvoltages produced by rapid modifications in the network structure such as opening of protective devices or the switching on of capacitive circuits.

Overvoltage may cause over stress on insulation, problems of voltage instability, demand for reactive power, etc.

6.8 Discussion

The power quality issues described in the previous sections are serious in the present competitive power market where a consumer has the freedom to choose a supply out of many. This has made technologists to throw more and more light on detection of these problems and their remedial measure.

References

[1] Bollen, M.H.J.: Understanding power quality problems. IEEE Press Series on Power Engineering, New York (2000)

[2] Fuchs, E.F., Masoum, M.A.S.: Power quality in power systems and electrical machines, AP, ISBN 978-0-12-369536

[3] IEEE Recommended Practice for Monitoring Electric Power Quality. IEEE Standard, pp. 1159–1995 (1995)

[4] Zhang, L.D., Bollen, M.H.J.: A method for characterization of three-phase unbalanced dips (sags) from recorded voltage wave shapes. Proceedings of International Telecommunications Energy Conference (INTELEC), Copenhagen (1999)

[5] Zhang, L.D., Bollen, M.H.J.: A method for characterizing unbalanced voltage dips (sags) with symmetrical components. IEEE Power Eng. Rev. **18**, 50–52 (1998)

Chapter 7
DC Offset, Electric Noise, Voltage Fluctuation, Flicker and Power Frequency Variation

Abstract This chapter deals with some of the power system disturbances like, DC offset, voltage fluctuation, electric noise, flicker and power frequency variation. The probable causes and effects of these disturbances are explained.

7.1 Introduction

Waveform distortion in power system refers to steady-state deviation from a sine wave of power frequency. Main cause of waveform distortion includes DC offset, harmonics, inter-harmonics, notching, and electric noise. Voltage fluctuation, flicker and power frequency variation are other important reason for waveform distortion [1–4]. However monitoring of these disturbances is not too difficult which mainly include monitoring of rms value and/or power frequency except for noise.

7.2 DC Offset

DC offset is the presence of a DC current and/or voltage component in an AC system. Expansion of ac wave by Fourier series is used for analysis of DC offset [5]. Main causes of DC offset in power systems are:

- Operation of rectifiers and other electronic switching devices, and
- Geomagnetic disturbances causing GICs.

The main detrimental effects of DC offset in alternating networks are:

- half-cycle saturation of transformer core
- generation of even harmonics in addition to odd harmonics
- additional heating in appliances which may lead to a decrease of the lifetime of transformers, rotating machines, and electromagnetic devices
- electrolytic erosion of grounding electrodes and other connectors.

S. Chattopadhyay et al., *Electric Power Quality,* Power Systems,
DOI 10.1007/978-94-007-0635-4_7,

For suppression of DC currents due to rectifiers and geomagnetically induced currents, three-limb transformers with a relatively large air gap between core and tank are used.

7.3 Electric Noise

Electric noise is unwanted electrical signals present in power system having broadband spectral content lower than 200 kHz superimposed with the power system voltage or current. They may be observed in phase conductors, or neutral conductors or signal lines. Main causes of electric noise are:

- faulty connections in transmission or distribution systems
- arc furnaces
- electrical furnaces
- power electronic devices
- control circuits
- welding equipment
- loads with solid-state rectifiers
- improper grounding
- turning off capacitor banks
- adjustable-speed drives
- corona
- interference with communication circuits

Magnitude of noise is normally very small as compared with power signal. However, electric noise creates problems in operation of electronic devices such as microcomputers and programmable controllers.

Electric noise is mitigated by using filters, line conditioners, and dedicated lines or transformers.

7.4 Voltage Fluctuation

Voltage fluctuations are variations of voltage, magnitude of which does not normally exceed specified voltage ranges. According to ANSI C84.1-1982 the range of fluctuation of voltage magnitude is from 0.9 to 1.1 pu. Based on the nature of voltage fluctuations, the disturbance is classified as:

- step-voltage changes which may be regular or irregular with respect to time
- cyclic voltage fluctuation with respect to time
- random voltage fluctuation with respect to time

The main causes of voltage fluctuation are:

- pulsed-power output
- resistance welders

- start-up of drives
- arc furnaces
- drives with rapidly changing loads or load impedance
- rolling mills

Voltage fluctuations results in:

- degradation of the performance of the equipment
- instability of the internal voltages and currents of electronic equipment
- problem in reactive power compensation

(Note: voltage fluctuations less than 10% do not create severe problem in electronic equipment.)

7.5 Flicker

Flicker refers to "continuous and rapid variations in the load current magnitude which causes voltage variations" [4]. Flicker is one type of voltage fluctuation. The term flicker is useful in illumination technology where the effect of voltage fluctuation on lamps is perceived as flicker by the human eye. Operation of an arc furnace is one of the most common causes of the flicker voltage fluctuations in utility transmission and distribution systems.

7.6 Power Frequency Variations

Power frequency variation refers to the deviation of the power system fundamental frequency from its specified nominal value (e.g., 50 or 60 Hz). Main cause of power frequency variation is the difference between active power demand and power generation. It depends on the generator load characteristics and control circuit for automatic load frequency control (ALFC) which includes governor control as well as turbine control. The frequency variation is controlled by proper balance between generation and demand (load), i.e., active power demand and mechanical energy supplied to the prime mover.

7.7 Discussion

The number of power system disturbances described in the previous sections has increased to a great extent which is a serious concern specially with respect to stability and proper functioning of measuring instruments, control devices and protection equipment. This has thrown a big challenge to the power engineers to detect, isolate and find out the location of source of such disturbance. Adding with this, solution and prediction of such fault will be very much welcome.

References

[1] Huang, S-J., Yang, T-M., Huang, J-T.: FPGA realization of wavelet transform for detection of electric power system disturbances. IEEE Trans. Power Deliv. **17**(2), 388–394 (2002)

[2] Bollen, M.H.J.: Understanding power quality problems. IEEE Press Series on Power Engineering, New York (2000)

[3] Fuchs, E.F., Masoum, M.A.S.: Power Quality in Power Systems and Electrical Machines, AP, ISBN 978-0-12-369536

[4] Larsson, A.: Flicker and slow voltage variations from wind power. In Proceedings of International Conference Harmonics and Quality of Power, Las Vegas, pp. 270–275 (1996)

[5] Santoso, S., Powers, E.J., Grady, W.M., Parsons, A.C.: Characterization of distribution power quality events with fourier and wavelet transform. IEEE Trans. Power Deliv. **15**(1), 247–256 (2000)

Chapter 8
Unbalance Assessment Using Sequence Components

Abstract This chapter starts with an introduction of unbalance assessment. It then describes sequence components in three phase system. Unbalance is assessed using sequence components. A case study has been shown on single phasing of an induction machine. Generation of sequence components along with their phase shift has been mathematically explained.

8.1 Introduction

During unbalance of a power system, negative and zero sequence components appear in the system. For this reason, traditional unbalance assessment is based on analysis of sequence components generated during unbalance situation. First, all sequence components are determined, then they are added vectorially and unbalance is assessed [1–3]. In the following sections, first different sequence components are described, then a case study has been done for unbalance assessment. In some research work, characterization of unbalance and sag has been done based on the assessment of sequence components [4, 6]. Then a new definition of unbalance has been described with respect to the phase shift occurred during unbalance [5].

8.2 Sequence Component

In generalized n phase system, n number of sequence-components appears during unbalance. In three phase system, three sequence components appear [3]; they are

1. positive sequence component
2. negative sequence component
3. zero sequence component

All these three sequence components are described in the following sections.

S. Chattopadhyay et al., *Electric Power Quality,* Power Systems,
DOI 10.1007/978-94-007-0635-4_8,

8.2.1 Positive Sequence Current and Voltage Components

Rotation and sequence of positive sequence components are same as that of normal phase signals at balanced condition. The positive sequence current and voltage components for three phases namely R, Y and B are denoted by

Current: I_{R1}, I_{Y1} and I_{B1}
Voltage: V_{R1}, V_{Y1} and V_{B1}

The relations between them are given by

$$\begin{aligned} I_{R1} &= I_{Y1}\angle 120^\circ \\ I_{Y1} &= I_{B1}\angle 120^\circ \\ I_{B1} &= I_{R1}\angle 120^\circ \end{aligned} \tag{8.1}$$

$$\begin{aligned} V_{R1} &= V_{Y1}\angle 120^\circ \\ V_{Y1} &= V_{B1}\angle 120^\circ \\ V_{B1} &= V_{R1}\angle 120^\circ \end{aligned} \tag{8.2}$$

Both the rotation and sequence are in anti-clockwise for positive sequence. So the positive sequence components of currents and voltages for three-phases are equal in magnitude and displaced by phase of 120° as shown in Figs. 8.1 and 8.2.

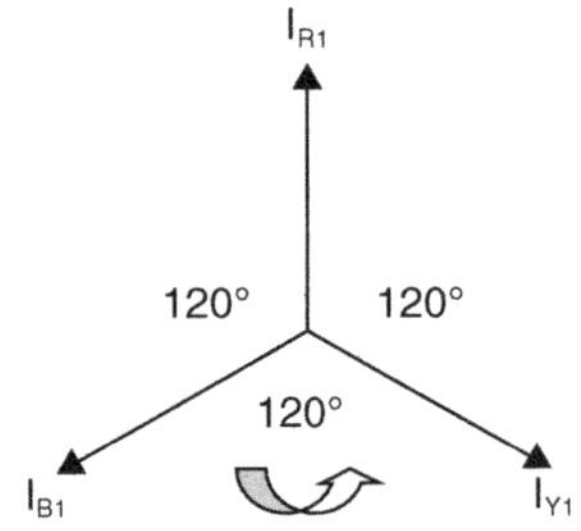

Fig. 8.1 Phasor orientation of positive sequence current components

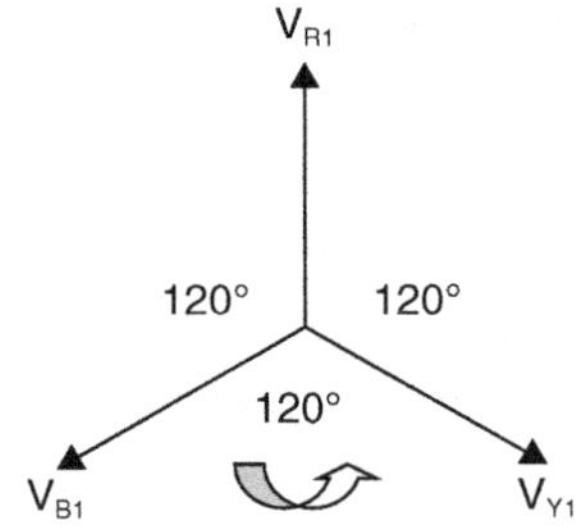

Fig. 8.2 Phasor orientation of positive sequence voltage components

8.2.2 Negative Sequence Current and Voltage Components

The negative sequence current and voltage components for the three phases namely R, Y and B are denoted by

Current: I_{R2}, I_{Y2} and I_{B2}
Voltage: V_{R2}, V_{Y2} and V_{B2}

The relations between them are given by

$$\begin{aligned} I_{R2} &= I_{B2}\angle 120^\circ \\ I_{B2} &= I_{Y2}\angle 120^\circ \\ I_{Y2} &= I_{R2}\angle 120^\circ \end{aligned} \tag{8.3}$$

$$\begin{aligned} V_{R2} &= V_{B2}\angle 120^\circ \\ V_{B2} &= V_{Y2}\angle 120^\circ \\ V_{Y2} &= V_{R2}\angle 120^\circ \end{aligned} \tag{8.4}$$

The rotation is considered in anti-clockwise direction whereas the sequence is in clock wise direction. So the negative sequence components of the three phases are equal in magnitude and with 120° phase displacement as shown in Figs. 8.3 and 8.4.

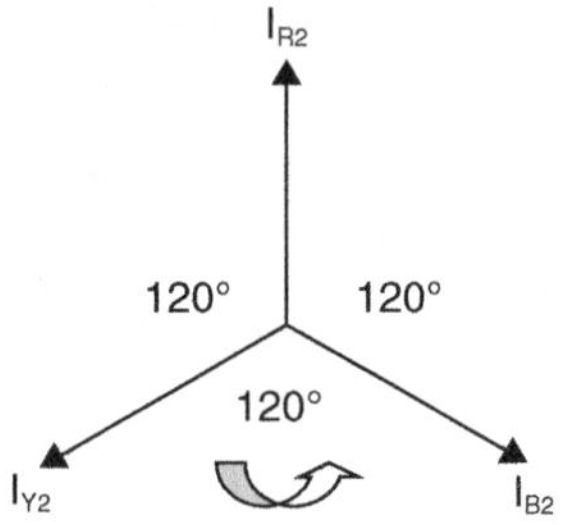

Fig. 8.3 Phasor orientation of negative sequence current components

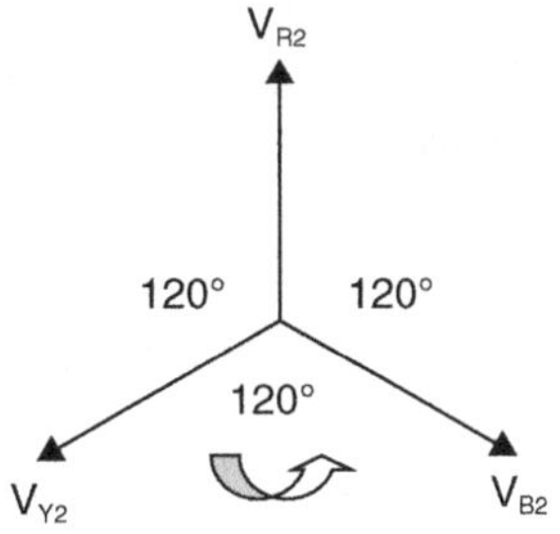

Fig. 8.4 Phasor orientation of negative sequence voltage components

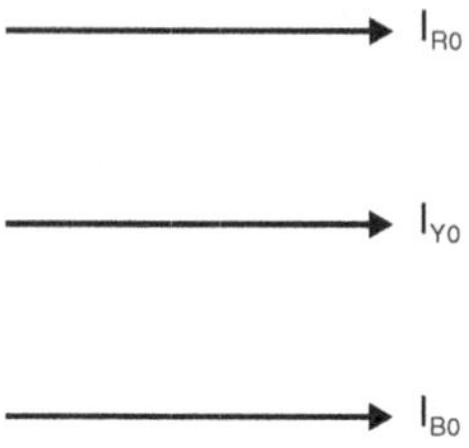

Fig. 8.5 Phasor orientation of zero sequence current components

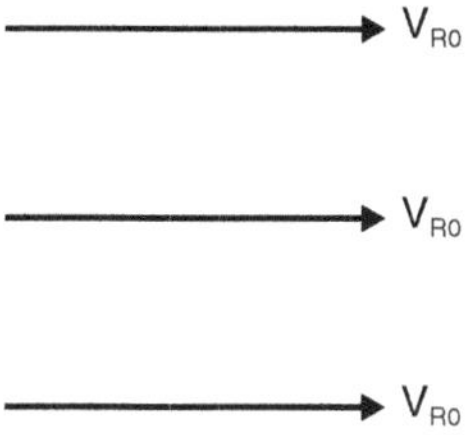

Fig. 8.6 Phasor orientation of zero sequence voltage components

8.2.3 Zero Sequence Current and Voltage Components

The zero sequence current and voltage components for the three phases namely R, Y and B are denoted by

Current: I_{R0}, I_{Y0} and I_{B0}
Voltage: V_{R0}, V_{Y0} and V_{B0}

The relations between them are given by

$$\begin{aligned} I_{R0} &= I_{Y0}\angle 0^\circ \\ I_{Y0} &= I_{B0}\angle 0^\circ \\ I_{B0} &= I_{R0}\angle 0^\circ \end{aligned} \tag{8.5}$$

$$\begin{aligned} V_{R0} &= V_{Y0}\angle 0^\circ \\ V_{Y0} &= V_{B0}\angle 0^\circ \\ V_{B0} &= V_{R0}\angle 0^\circ \end{aligned} \tag{8.6}$$

The zero sequence components of the three phases are equal in magnitude and with 0° phase displacement as shown in Figs. 8.5 and 8.6.

8.3 Phase Currents and Voltages

8.3.1 Balanced System

In a balanced three phase-supply, the negative and zero sequence components are absent. Therefore three phase current or voltages are actually the positive sequence

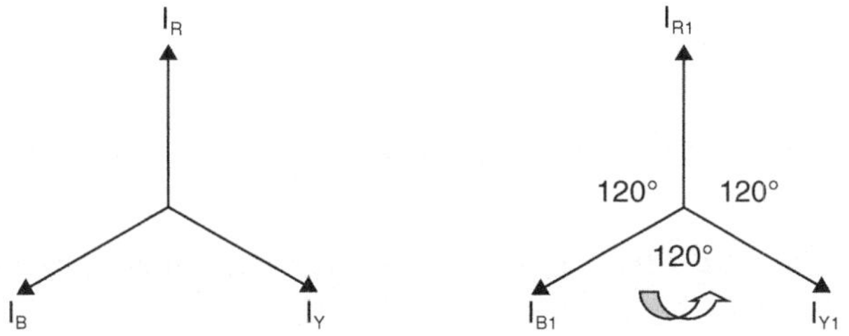

Fig. 8.7 Phasor diagram of phase currents and positive sequence currents

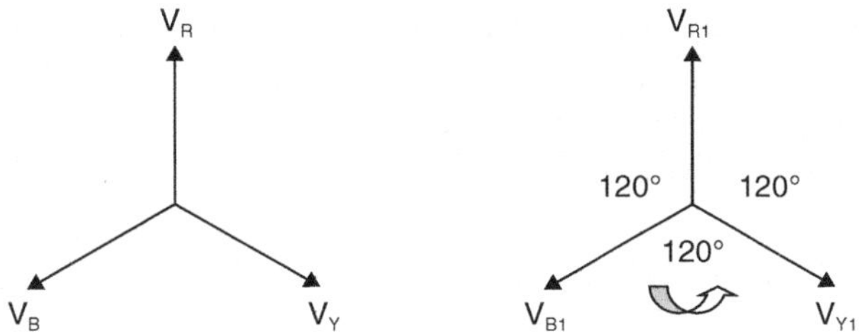

Fig. 8.8 Phasor diagram of phase voltages and positive sequence voltages

current and voltage components given by

$$I_R = I_{R1} \tag{8.7}$$

$$I_Y = I_{Y1} \tag{8.8}$$

$$I_B = I_{B1} \tag{8.9}$$

$$V_R = V_{R1} \tag{8.10}$$

$$V_Y = V_{Y1} \tag{8.11}$$

$$V_B = V_{B1} \tag{8.12}$$

Under balanced operating condition of the three-phase phase currents and phase voltages are ideally equal in magnitude and at a phase difference of 120° and they consist of only positive sequence components as shown in Figs. 8.7 and 8.8.

8.3.2 Unbalanced System

During unbalance, the three phase currents and voltages consist of positive, negative and zero sequence components given by

$$I_R = I_{R0} + I_{R1} + I_{R2} \tag{8.13}$$

$$I_Y = I_{Y0} + I_{Y1} + I_{Y2} \tag{8.14}$$

$$I_B = I_{B0} + I_{B1} + I_{B2} \tag{8.15}$$

$$V_R = V_{R0} + V_{R1} + V_{R2} \tag{8.16}$$

$$V_Y = V_{Y0} + V_{Y1} + V_{Y2} \tag{8.17}$$

$$V_B = V_{B0} + V_{B1} + V_{B2} \tag{8.18}$$

8.4 'a' Operator and Angle Representation in Complex Plane

The operator 'a' is a mathematical operator used to represent angle in complex plane. The operator is very useful in representation of phasor of poly-phase system. For n-phase system 'a' is defined as

$$a^n = 1 \tag{8.19}$$

For three phase system,

$$a^3 = 1 \tag{8.20}$$

Roots of (8.20) are 1, $\left(-\frac{1}{2} + j\frac{\sqrt{3}}{2}\right)$ and $\left(-\frac{1}{2} - j\frac{\sqrt{3}}{2}\right)$. The complex roots are represented by 'a' and 'a^2' as follows

$$a = e^{j\frac{2\pi}{3}} = \left(-\frac{1}{2} + j\frac{\sqrt{3}}{2}\right) \tag{8.21}$$

and

$$a^2 = e^{j\frac{4\pi}{3}} = \left(-\frac{1}{2} - j\frac{\sqrt{3}}{2}\right) \tag{8.22}$$

Therefore, sum of the roots are given by

$$1 + a + a^2 = 0 \tag{8.23}$$

Roots are used for angle representation as,

$$\angle 1 = 0^\circ \tag{8.24}$$

$$\angle\left(-\frac{1}{2} + j\frac{\sqrt{3}}{2}\right) = 120^\circ = a \tag{8.25}$$

$$\angle\left(-\frac{1}{2} - j\frac{\sqrt{3}}{2}\right) = 240^\circ = a^2 \tag{8.26}$$

$$\angle -1 = 180^\circ \tag{8.27}$$

$$\angle -\left(-\frac{1}{2} + j\frac{\sqrt{3}}{2}\right) = 300^\circ = -a \tag{8.28}$$

Fig. 8.9 Loci of 1, a, a^2, −1, −a. $-a^2$ in real and imaginary plane

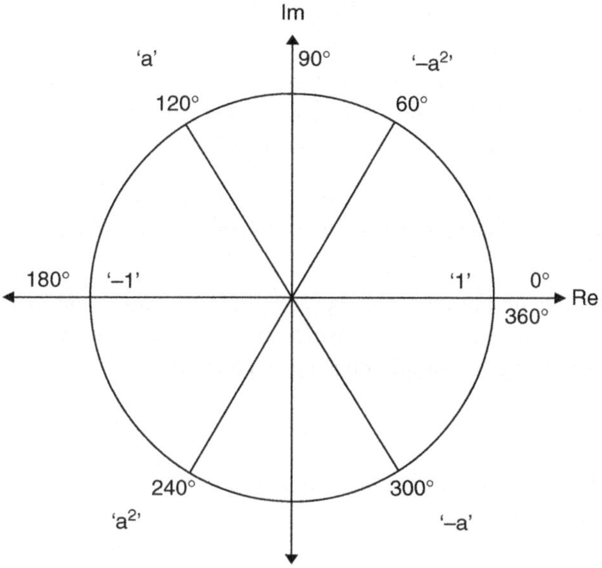

$$\angle - \left(-\frac{1}{2} - j\frac{\sqrt{3}}{2}\right) = -60^\circ = -a^2 \tag{8.29}$$

In real and imaginary plane, the loci of 1, a, a^2, −1, −a. $-a^2$ are shown in Fig. 8.9.

8.5 Currents and Voltages in Terms of Sequence Components with ‘a’ Operator

Considering $I_{R0} = I_{Y0} = I_{B0}$, $I_{R1} = I_{Y1} = I_{B1}$ and $I_{R2} = I_{Y2} = I_{B2}$, with the help of ‘a’ operator, (8.13), (8.14) and (8.15) can be written as

$$I_R = I_{R0} + I_{R1} + I_{R2} \tag{8.30}$$

$$I_Y = I_{R0} + a^2 I_{R1} + a I_{R2} \tag{8.31}$$

$$I_B = I_{R0} + a I_{R1} + a^2 I_{R2} \tag{8.32}$$

In matrix form, (8.30), (8.31) and (8.32) can be written as

$$\begin{bmatrix} I_R \\ I_Y \\ I_B \end{bmatrix} = \begin{bmatrix} 1 & 1 & 1 \\ 1 & a^2 & a \\ 1 & a & a^2 \end{bmatrix} \begin{bmatrix} I_{R0} \\ I_{R1} \\ I_{R2} \end{bmatrix} \tag{8.33}$$

Similarly, (8.16), (8.17) and (8.18) can be written as

$$\begin{bmatrix} V_R \\ V_Y \\ V_B \end{bmatrix} = \begin{bmatrix} 1 & 1 & 1 \\ 1 & a^2 & a \\ 1 & a & a^2 \end{bmatrix} \begin{bmatrix} V_{R0} \\ V_{R1} \\ V_{R2} \end{bmatrix} \tag{8.34}$$

8.6 Case Study on Unbalance

For study of unbalance in terms of sequence component, single phasing phenomena of an induction motor is considered [5]. Mathematical relations for line currents, sequence components and phase relations are established during single phasing.

8.6.1 Single Phasing in Induction Motor

Let us first assume that before single phasing of an induction motor all three phases were healthy. Then consider single phasing occurs due the absence of phase B. Therefore, for single phasing in phase B,

$$I_B = 0 \tag{8.35}$$

Similarly, for single phasing in phase Y,

$$I_Y = 0 \tag{8.36}$$

and for single phasing in phase R,

$$I_R = 0 \tag{8.37}$$

8.6.2 Line Currents during Single Phasing

For single phasing at phase B, when $I_{R0} = 0$, (8.30), (8.31) and (8.32) can be written as

$$I_R = I_{R1} + I_{R2} \tag{8.38}$$

$$I_Y = a^2 I_{R1} + a I_{R2} \tag{8.39}$$

$$0 = a I_{R1} + a^2 I_{R2} \tag{8.40}$$

From (8.23),

$$a + a^2 = -1 \tag{8.41}$$

From (8.38) and (8.41),

$$I_R = I_{R1} + I_{R2} = -1 \times -1 \times (I_{R1} + I_{R2}) = -1 \times \left(a + a^2\right) \times (I_{R1} + I_{R2})$$

$$I_R = -\left(a I_{R1} + a^2 I_{R2} + a I_{R2} + a^2 I_{R1}\right) \tag{8.42}$$

From (8.42) and (8.40),

$$I_R = -\left(a I_{R2} + a^2 I_{R1}\right) \tag{8.43}$$

From (8.39) and (8.43)

$$I_R = -I_Y \tag{8.44}$$

It is proved from (8.44) that, during single phasing at phase B, line current in phase R and Y are opposite in phase. Similarly, for single phasing at phase R

$$I_Y = -I_B \tag{8.45}$$

and for single phasing at phase Y

$$I_B = -I_R \tag{8.46}$$

So, it is interesting to note that at single phasing of a particular phase in a three phase system the other two phases of the system are exactly opposite in phase to each other i.e. at a phase difference of 180° between them.

8.6.3 Sequence Components in Single Phasing

From (8.40),

$$0 = aI_{R1} + a^2 I_{R2} \tag{8.47}$$

or,

$$I_{R1} = -aI_{R2}$$

or,

$$a^2 I_{R1} = -a^3 I_{R2} = -I_{R2} \tag{8.48}$$

$$\begin{aligned} I_{R2} &= -a^2 I_{R1} \\ &= I_{R1} \angle -240^\circ \\ &= I_{R1} \angle 60^\circ \end{aligned} \tag{8.49}$$

$$\therefore I_{R2} = I_{R1} \angle 60^\circ \tag{8.50}$$

$$\begin{aligned} I_{Y1} &= a^2 I_{R1} \\ &= a^2 \times -aI_{R2} \\ &= -a^3 I_{R2} \\ &= -I_{R2} \\ &= I_{R2} \angle 180^\circ \end{aligned} \tag{8.51}$$

$$I_{Y2} = aI_{R2} = I_{R2} \angle 120^\circ \tag{8.52}$$

$$\therefore I_{Y1} = I_{Y2} \angle 60^\circ \tag{8.53}$$

$$
\begin{aligned}
I_{B1} &= aI_{R1} \\
&= a \times -aI_{R2} \\
&= -a^2 I_{R2} \\
&= I_{R2}\angle - 240^\circ \\
&= I_{R2}\angle 60^\circ
\end{aligned} \tag{8.54}
$$

$$
I_{B2} = a^2 I_{R2} = I_{R2}\angle 240^\circ \tag{8.55}
$$

$$
\therefore I_{B1} = I_{B2}\angle 180^\circ \tag{8.56}
$$

Then, from (8.50), (8.53) and (8.56),

$$
\therefore \angle\,(I_{R1}, I_{R2}) = \angle 60^\circ \tag{8.57}
$$

$$
\therefore \angle\,(I_{Y1}, I_{Y2}) = \angle 60^\circ \tag{8.58}
$$

$$
\therefore \angle\,(I_{B1}, I_{B2}) = \angle 180^\circ \tag{8.59}
$$

Similarly, single phasing at phase R gives

$$
I_{R1} = I_{R2}\angle 180^\circ \tag{8.60}
$$

$$
I_{Y2} = I_{Y1}\angle 60^\circ \tag{8.61}
$$

$$
I_{B1} = I_{B2}\angle 60^\circ \tag{8.62}
$$

For single phasing at phase Y

$$
I_{R1} = I_{R2}\angle 60^\circ \tag{8.63}
$$

$$
I_{Y1} = I_{Y2}\angle 180^\circ \tag{8.64}
$$

$$
I_{B2} = I_{B1}\angle 60^\circ \tag{8.65}
$$

From the above derivations at single phasing it is observed that the negative sequence components are at an angle of 60° to that of the positive sequence components in healthy lines and 180° out of phase to that of positive sequence components at the line where single phasing has taken place.

If current in one phase of the motor gradually decreases, causing unbalance in all the phases, the negative sequence components will start generating. The relative phase orientations of the negative sequence components with that of positive sequence components depend on which phase the unbalance occur. The magnitude of the negative sequence components increases proportionally as the degree of unbalance increases and at maximum unbalance i.e. at single phasing, when current in one phase becomes absent, the magnitude of negative sequence components are equal to that of the positive sequence components. Figures 8.10a–c show the sequence components during single phasing at phase R, Y and B respectively.

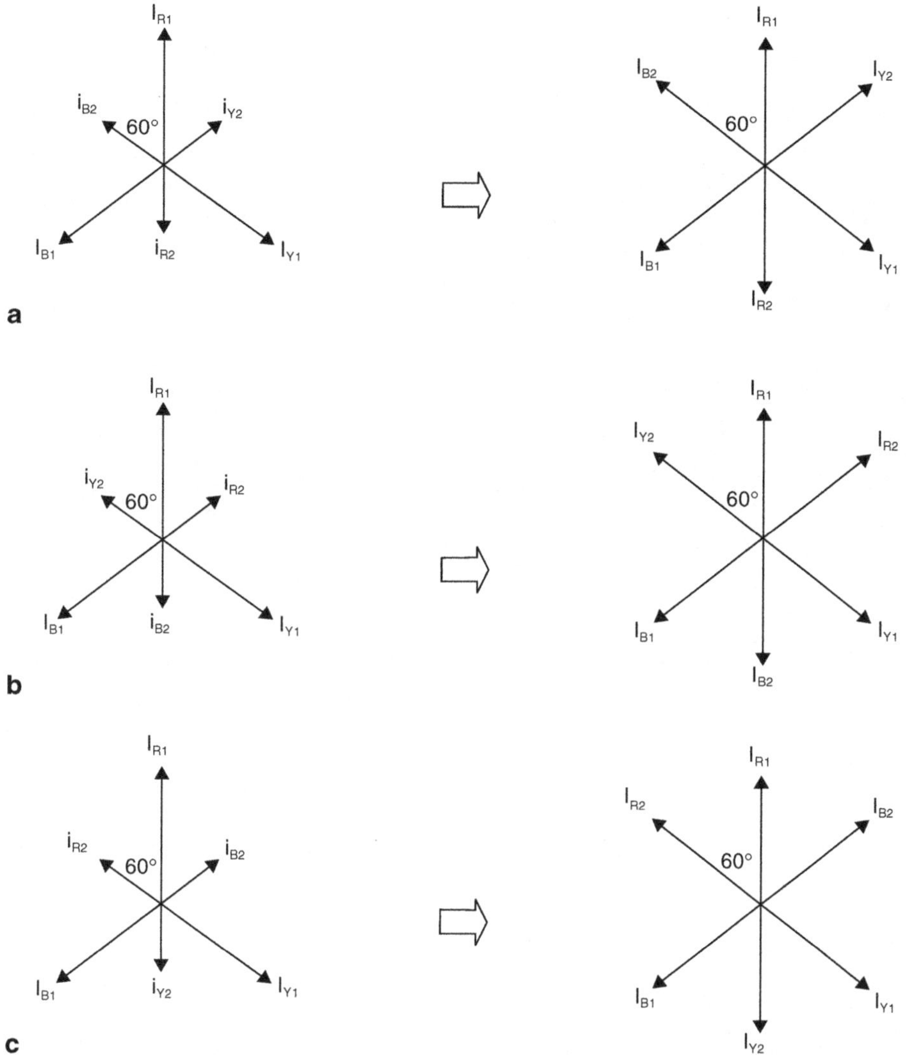

Fig. 8.10 Phasor diagram of negative sequence currents and space orientation between positive and negative sequence currents for **a** unbalance in phase R, **b** unbalance in phase Y and **c** unbalance in phase B

During unbalance, currents are given by the sum of sequence components. So during unbalance or single phasing in phase R, Y and B respectively, the line currents can be drawn as shown in Figs. 8.11a–c respectively.

From the above diagrams, it is clearly seen that for single phasing at phase B, the line current vector of R phase rotates through an angle 30° in anti-clockwise direction, whereas that of the Y phase rotates by the same angle in clockwise direction resulting

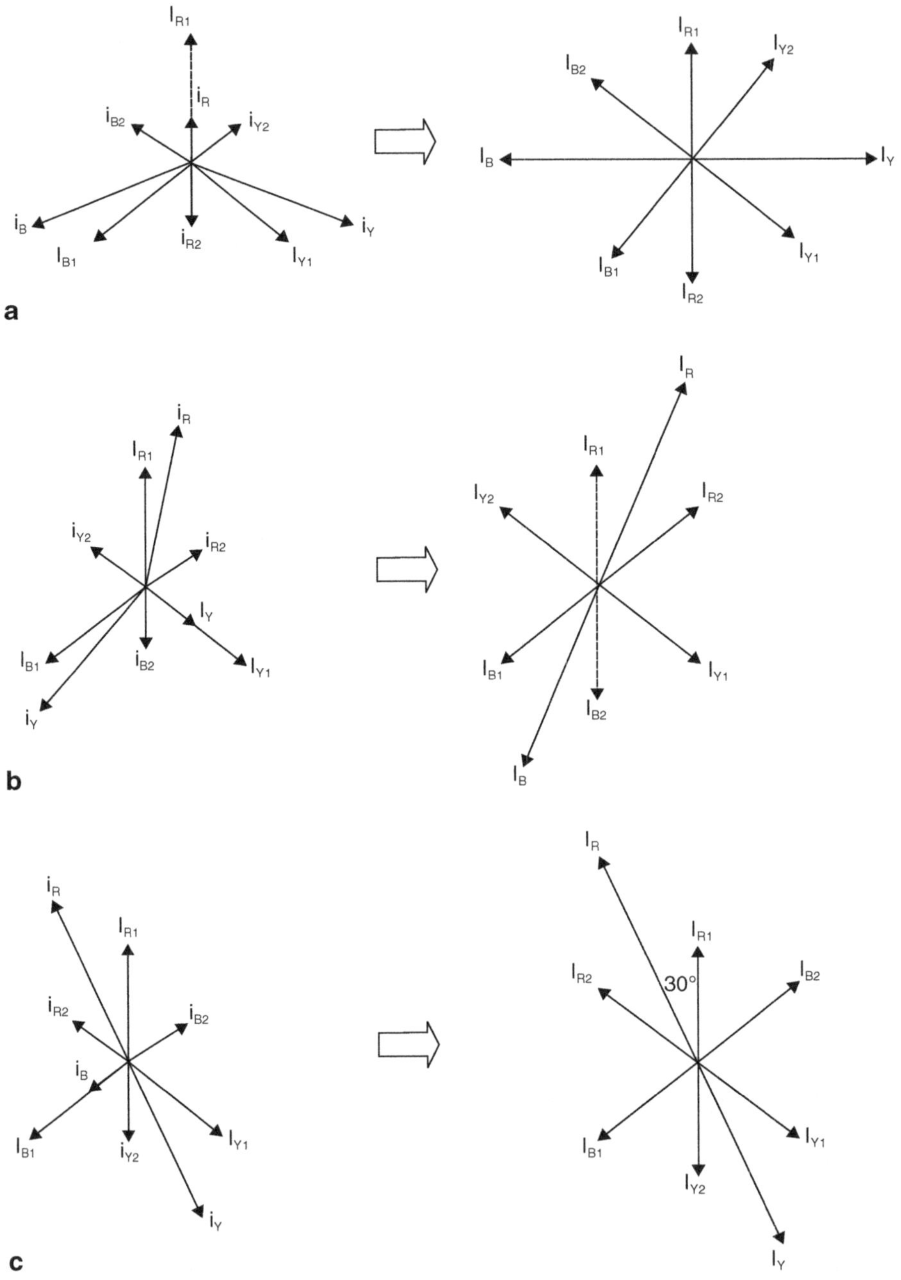

Fig. 8.11 Phasor diagram of change in stator current as the degree of unbalance increases for unbalance in **a** phase R, **b** phase Y and **c** phase

in a net phase difference of 180° in place of 120° between R and Y phases. This phase difference increase is related to degree of unbalance. So characteristics between the percentage change in current (i.e. unbalance) in phase B and change in the phase difference between the other two line currents can be obtained.

8.6.4 Line Currents and Sequence Components

From Fig. 8.11a, for single phasing at phase R

$$
\begin{aligned}
I_R &= \sqrt{I_{R1}^2 + I_{R2}^2 + 2I_{R1}I_{R2}\cos 180^\circ} \\
&= \sqrt{I_{R1}^2 + I_{R1}^2 + 2I_{R1}I_{R1}\cos 180^\circ} \\
I_R &= 0 \qquad (8.66) \\
I_Y &= \sqrt{I_{Y1}^2 + I_{Y2}^2 + 2I_{Y1}I_{Y2}\cos 60^\circ} \\
&= \sqrt{I_{Y1}^2 + I_{Y1}^2 + 2I_{Y1}I_{Y1}\cos 60^\circ} \\
I_Y &= \sqrt{3}I_{Y1} \qquad (8.67) \\
I_B &= \sqrt{I_{B1}^2 + I_{B2}^2 + 2I_{B1}I_{B2}\cos 60^\circ} \\
&= \sqrt{I_{B1}^2 + I_{B1}^2 + 2I_{B1}I_{B1}\cos 60^\circ} \\
I_B &= \sqrt{3}I_{B1} \qquad (8.68)
\end{aligned}
$$

$$
\tan\theta_B = \frac{I_{B2}\sin 60^\circ}{(I_{B1} + I_{B2}\cos 60^\circ)} = \frac{I_{B1}\sin 60^\circ}{(I_{B1} + I_{B1}\cos 60^\circ)} = \frac{1}{\sqrt{3}} \qquad (8.69)
$$

$$
\therefore \theta_B = 30^\circ \qquad (8.70)
$$

$$
\tan\theta_y = \frac{I_{Y2}\sin 60^\circ}{(I_{Y1} + I_{Y2}\cos 60^\circ)} = \frac{I_{Y1}\sin 60^\circ}{(I_{Y1} + I_{Y1}\cos 60^\circ)} = \frac{1}{\sqrt{3}} \qquad (8.71)
$$

$$
\therefore \theta_Y = 30^\circ \qquad (8.72)
$$

Before, single phasing

$$
\angle\,(I_Y, I_B) = 120^\circ \qquad (8.73)
$$

After, single phasing

$$
\angle\left(I_y, I_B\right) = 120^\circ + 30^\circ + 30^\circ = 180^\circ \qquad (8.74)
$$

From Fig. 8.11b, for single phasing at phase Y

$$I_R = \sqrt{I_{R1}^2 + I_{R2}^2 + 2I_{R1}I_{R2}\cos 60^\circ}$$
$$= \sqrt{I_{R1}^2 + I_{R1}^2 + 2I_{R1}I_{R1}\cos 60^\circ}$$
$$I_R = \sqrt{3}I_{R1} \tag{8.75}$$

$$I_Y = \sqrt{I_{Y1}^2 + I_{Y2}^2 + 2I_{Y1}I_{Y2}\cos 180^\circ}$$
$$= \sqrt{I_{Y1}^2 + I_{Y1}^2 + 2I_{Y1}I_{Y1}\cos 180^\circ}$$
$$I_Y = 0 \tag{8.76}$$

$$I_B = \sqrt{I_{B1}^2 + I_{B2}^2 + 2I_{B1}I_{B2}\cos 60^\circ}$$
$$= \sqrt{I_{B1}^2 + I_{B1}^2 + 2I_{B1}I_{B1}\cos 60^\circ}$$
$$I_B = \sqrt{3}I_{B1} \tag{8.77}$$

$$\tan\theta_R = \frac{I_{R2}\sin 60^\circ}{(I_{R1} + I_{R2}\cos 60^\circ)} = \frac{I_{R1}\sin 60^\circ}{(I_{R1} + I_{R1}\cos 60^\circ)} = \frac{1}{\sqrt{3}} \tag{8.78}$$

$$\therefore \theta_R = 30^\circ \tag{8.79}$$

$$\tan\theta_B = \frac{I_{B2}\sin 60^\circ}{(I_{B1} + I_{B2}\cos 60^\circ)} = \frac{I_{B1}\sin 60^\circ}{(I_{B1} + I_{B1}\cos 60^\circ)} = \frac{1}{\sqrt{3}} \tag{8.80}$$

$$\therefore \theta_B = 30^\circ \tag{8.81}$$

Before, single phasing

$$\angle(I_B, I_R) = 120^\circ \tag{8.82}$$

After, single phasing

$$\angle(I_B, I_R) = 120^\circ + 30^\circ + 30^\circ = 180^\circ \tag{8.83}$$

From Fig. 8.11c, for single phasing at phase B

$$I_R = \sqrt{I_{R1}^2 + I_{R2}^2 + 2I_{R1}I_{R2}\cos 60^\circ}$$
$$= \sqrt{I_{R1}^2 + I_{R1}^2 + 2I_{R1}I_{R1}\cos 60^\circ}$$

$$I_R = \sqrt{3} I_{R1} \tag{8.84}$$

$$I_Y = \sqrt{I_{Y1}^2 + I_{Y2}^2 + 2I_{Y1}I_{Y2}\cos 60^\circ}$$

$$= \sqrt{I_{Y1}^2 + I_{Y1}^2 + 2I_{Y1}I_{Y1}\cos 60^\circ}$$

$$I_Y = \sqrt{3} I_{R1} \tag{8.85}$$

$$I_B = \sqrt{I_{B1}^2 + I_{B2}^2 + 2I_{B1}I_{B2}\cos 180^\circ}$$

$$= \sqrt{I_{B1}^2 + I_{B1}^2 + 2I_{B1}I_{B1}\cos 180^\circ}$$

$$I_B = 0 \tag{8.86}$$

$$\tan\theta_R = \frac{I_{R2}\sin 60^\circ}{(I_{R1} + I_{R2}\cos 60^\circ)} = \frac{I_{R1}\sin 60^\circ}{(I_{R1} + I_{R1}\cos 60^\circ)} = \frac{1}{\sqrt{3}} \tag{8.87}$$

$$\therefore \theta_R = 30^\circ \tag{8.88}$$

$$\tan\theta_y = \frac{I_{Y2}\sin 60^\circ}{(I_{Y1} + I_{Y2}\cos 60^\circ)} = \frac{I_{Y1}\sin 60^\circ}{(I_{Y1} + I_{Y1}\cos 60^\circ)} = \frac{1}{\sqrt{3}} \tag{8.89}$$

$$\therefore \theta_Y = 30^\circ \tag{8.90}$$

Before, single phasing

$$\angle(I_R, I_Y) = 120^\circ \tag{8.91}$$

After, single phasing

$$\angle(I_R, I_Y) = 120^\circ + 30^\circ + 30^\circ = 180^\circ \tag{8.92}$$

8.7 Definition of Unbalance: An Alternate Approach

In the previous sections it has been observed that, as the current in any one of the phases is limited or decreased, causing unbalance, currents in the other two healthy phases increase in a linear proportion. Along with the increase of currents in the healthy phases, their phase difference also changes (e.g.: it increases from 120° to 180° in single phasing of an induction machine). Based on this phenomenon, a novel concept is introduced where, "**percentage of unbalance**" is defined in terms of phase shift [5]. By this definition, percentage of unbalance is

$$\text{Percentage Unbalance} = \text{f}\left(\frac{\Theta - \theta}{\Theta} \times 100\right)\% \tag{8.93}$$

where, Θ is the phase difference at balance condition (i.e., equal to 120° for three phase system) and θ is the phase difference during unbalance condition.

References

[1] Ghijselen, J.A.L., Van den Bossche, A.P.M.: Exact voltage unbalance assessment without phase measurement. IEEE Trans. Power Syst. **20**(1), 519–520 (2005)

[2] Leva, S., Moronado, A.P., Zaninelli, D.: Evaluation of the line voltage drops in presence of unbalance, harmonics and inter-harmonics: Theory and applications. IEEE Trans. Power Deliv. **20**(1), 390–396 (2005)

[3] Bollen, M.: Definitions of symmetrical coordinates applied to the solution of poly phase networks. IEEE Power Eng. Rev. **22**(11), 49–50 (2002)

[4] Bollen, M.H.J., Styvaktakis, E.: Characterization of three phase unbalance dips (as easy as one-two-three?). In Proceedings of 9th International IEEE Conference Harmonics Quality Power, vol. 1, pp. 81–86, Orlando, Oct. 1–4, 2000

[5] Chattopadhyay, S., Mitra, M., Sengupta, S.: A Novel Approach for Diagnosis of Current Unbalance at the Terminals of a Three-Phase Induction Motor. In Proceedings of MS2004-International Conference on Modeling and Simulation, AMSE (France), in Lyon, France, July 2004

[6] Zhang, L.D., Bollen, M.H.J.: A method for characterizing unbalanced voltage dips (sags) with symmetrical components. IEEE Power Eng. Rev. **18**, 50–52 (1998)

Chapter 9
Unbalance Assessment Using Feature Pattern Extraction Method

Abstract This chapter starts with an introduction of feature pattern extraction method. It then describes the theoretical background of the feature pattern extraction method for assessment of unbalance of voltage and current in a power system. The method is then applied on normalized waveform. Rule set is generated through computer simulation for assessment of unbalance using feature pattern extraction method. Merits and demerits of the methods are discussed.

9.1 Introduction

In Chap. 8, unbalance has been assessed by conventional analysis using sequence components. Here unbalance has been assessed by feature pattern extraction method (FPEM). Recently the method has been extensively used to solve different power system problems. Application of the method has some advantage of lesser data handling feature which reduces data space and time of execution. In the following section feature pattern extraction method has been described in short in the context of unbalance assessment. Normalized power system data are considered at different unbalance condition and the method is applied on these data. Rule set developed using FPEM has been presented for unbalance assessment.

9.2 Feature Pattern Extraction Method

Feature pattern extraction method (FPEM) is a branch of solving waveform related problems, where signatures of different power system parameters are captured; some specific patterns are formed with the signatures themselves or along with some other predefined reference signal, wherefrom main features of the patterns are extracted to analyse system performance or to solve problems [1, 2]. Extraction of the main feature reduces the amount of data to be stored and the data handling time and and makes the execution simple.

S. Chattopadhyay et al., *Electric Power Quality,* Power Systems,
DOI 10.1007/978-94-007-0635-4_9, © Springer Science+Business Media B.V. 2011

9.3 Unbalance and FPEM

For unbalance assessment using feature pattern extraction method (FPEM) of a power system, first electrical signals, mainly voltage and current are to be captured. Patterns are to be developed using these signals and the main features of the pattern are extracted wherefrom degree of unbalance is assessed [1–5]. A good quality power consists of balanced voltage and current waveforms free from harmonics. Voltages can be represented as

$$v_R(t) = V_R \sin \omega t \tag{9.1}$$

$$v_Y(t) = V_Y \sin (\omega t + 120^\circ) \tag{9.2}$$

$$v_B(t) = V_B \sin (\omega t + 240^\circ) \tag{9.3}$$

Similarly, currents can be represented as

$$i_R(t) = I_R \sin (\omega t + \phi) \tag{9.4}$$

$$i_Y(t) = I_Y \sin (\omega t + 120^\circ + \phi) \tag{9.5}$$

$$i_B(t) = I_B \sin (\omega t + 240^\circ + \phi) \tag{9.6}$$

For a balanced system,

$$V_R = V_Y = V_B \quad \text{and} \quad I_R = I_Y = I_B$$

From (9.1) and (9.2)

$$v_Y(t) = V_Y \sin \left(\frac{v_R(t)}{V_R} + 120^\circ \right) \tag{9.7}$$

From (9.2) and (9.3)

$$v_B(t) = V_B \sin \left(\frac{v_Y(t)}{V_Y} + 120^\circ \right) \tag{9.8}$$

From (9.3) and (9.1)

$$v_R(t) = V_R \sin \left(\frac{v_B(t)}{V_B} - 240^\circ \right) \tag{9.9}$$

Similarly, from (9.4), (9.5) and (9.6)

$$i_Y(t) = I_Y \sin \left(\frac{i_R(t)}{I_R} + 120^\circ \right) \tag{9.10}$$

$$i_B(t) = I_B \sin \left(\frac{i_Y(t)}{I_Y} + 120^\circ \right) \tag{9.11}$$

$$i_R(t) = I_R \sin\left(\frac{i_B(t)}{I_B} - 240^\circ\right) \tag{9.12}$$

(9.7), (9.8) and (9.9) represent three ellipses in voltage-voltage plane where any two phase-voltages are plotted along X and Y-axes. (9.10), (9.11) and (9.12) represent three ellipses in current-current plane where any two phase-currents are plotted along X and Y-axes. The major axis of the each ellipse makes +45° or −45° with X-axis. In an unbalanced system V_R, V_Y, V_B and I_R, I_Y, I_B will not be equal i.e.,

$$V_R \neq V_Y \neq V_B \quad \text{and} \quad I_R \neq I_Y \neq I_B$$

In an unbalanced system, normalized signals are used for analysis. Normalization is a process by which all signals are expressed in respect of one of these signals. In normalized form, voltages can be represented as,

$$v_R^N(t) = \frac{v_R(t)}{V_R} = \sin \omega t \tag{9.13}$$

$$v_Y^N(t) = \frac{v_Y(t)}{V_R} = \frac{V_Y}{V_R}\sin(\omega t + 120^\circ) \tag{9.14}$$

$$v_B^N(t) = \frac{v_B(t)}{V_R} = \frac{V_B}{V_R}\sin(\omega t + 240^\circ) \tag{9.15}$$

Similarly, currents can be represented, in normalized form, as

$$i_R^N(t) = \frac{i_R(t)}{I_R} = \sin(\omega t + \phi) \tag{9.16}$$

$$i_Y^N(t) = \frac{i_Y(t)}{I_R} = \frac{I_Y}{I_R}\sin(\omega t + 120^\circ) \tag{9.17}$$

$$i_B^N(t) = \frac{i_B(t)}{I_R} = \frac{I_B}{I_R}\sin(\omega t + 240^\circ) \tag{9.18}$$

Any two phase-voltages and any two line-currents of normalized data are plotted in voltage-voltage plane and current-current plane respectively. Patterns are formed in these planes. These are elliptical in nature. One such pattern formed by voltage signals has been shown in Fig. 9.1. The unbalance in a system changes amplitudes and phase angles of the signals. This results in change in the length of major and minor axes of the closed patterns.

Obviously, the patterns in voltage-voltage and current-current plane carry the information of unbalance in their shape. To extract the features from the patterns, i.e., to get back the information of voltage or current unbalance from the patterns, the following parameters are introduced:

$$\text{at } x = X_{MIN}(\text{point A}), \quad y = Y_1(\text{AE})$$

$$\text{at } x = X_{MAX}(\text{point C}), \quad y = Y_2(\text{CH})$$

$$Y = Y_1 \sim Y_2 \tag{9.19}$$

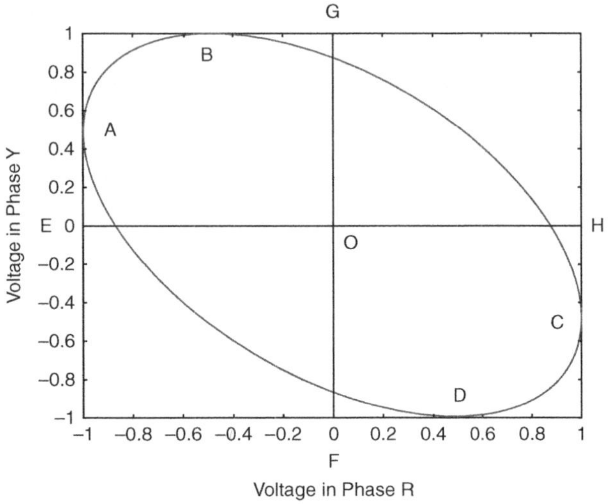

Fig. 9.1 Elliptical pattern in voltage-voltage plane

$$\begin{aligned} &\text{at } y = Y_{MIN}(\text{point D}), \quad x = X_1(FD) \\ &\text{at } y = Y_{MAX}(\text{point B}), \quad x = X_2(GB) \\ &X = X_1 \sim X_2 \end{aligned} \tag{9.20}$$

For assessment of voltage unbalance in three-phase system, any two phase-voltages are considered and plotted in voltage-voltage plane. For phase R and phase Y in voltage-voltage plane, phase R voltage is along x-axis and phase Y voltage is along y-axis. X and Y are calculated. They are denoted as X_{RY} and Y_{RY}.

Similarly, for phases Y and B and phases B and R as X_{YB} and Y_{YB} and X_{BR} and Y_{BR} are calculated from respective loops in voltage-voltage plane. Thus, in three phase system, two column matrices are obtained in voltage-voltage planes as follows:

$$[x_V] = \begin{bmatrix} X_{V_{RY}} \\ X_{V_{YB}} \\ X_{V_{BR}} \end{bmatrix} \tag{9.21}$$

$$[y_V] = \begin{bmatrix} Y_{V_{RY}} \\ Y_{V_{YB}} \\ Y_{V_{BR}} \end{bmatrix} \tag{9.22}$$

Note that these two matrices carry the features of pattern developed by two signals. Unbalance changes the dimensions of elliptical pattern. Changes in the length of major axis and minor axis of the elliptical patterns are reflected in the values of $[x_V]$ and $[y_V]$.

Similarly, for assessment of current unbalance in three-phase system, any two line currents are considered and plotted in current-current plane. For phase R and phase Y in current-current plane, Phase R current is considered along x-axis and phase Y current is considered along y-axis. X and Y are calculated as mentioned above. In same way, X and Y are calculated with Y and B phase currents and B and R phase current in current-current planes. Two matrices are formed for current unbalance assessment as follows

$$[x_I] = \begin{bmatrix} X_{I_{RY}} \\ X_{I_{YB}} \\ X_{I_{BR}} \end{bmatrix} \tag{9.23}$$

$$[y_I] = \begin{bmatrix} Y_{I_{RY}} \\ Y_{I_{YB}} \\ Y_{I_{BR}} \end{bmatrix} \tag{9.24}$$

Current unbalance changes the dimensions of elliptical pattern. Changes in the length of major axis and minor axis of the elliptical patterns are reflected in the values of $[x_V]$ and $[y_V]$. These two matrices are used for unbalance assessment [3, 4].

9.4 CMS Rule Set for Unbalance Assessment by FPEM

Matrices [x] and [y] for balanced and different unbalanced situations are determined using FPEM. To do this signals are then normalized with respect to R-phase voltage and current. In each case, any two voltage-signals and any two current-signals are plotted in voltage-voltage and current-current planes, respectively. Patterns are thus formed corresponding to phases R-Y, Y-B and B-R. Matrix [x] and matrix [y] for voltage and current are formed and taken into consideration for unbalance assessment.

Case I: Balanced System At first, let us consider a balanced system where amplitudes of all phase voltages are equal. Voltage signals are then normalized as shown in Fig. 9.2a. Patterns formed by those normalized signal will be as shown in Fig. 9.2b.

From patterns of Fig. 9.2b, $[x_V]$ and $[y_V]$ are calculated. These are presented in Table 9.1.

From Table 9.1, an inference can be considered as a mandatory rule for a system to be balanced. The rule is presented in Table 9.2.

Case II: Unbalance in Phase R Then unbalanced is introduced in R phase voltage by gradually increasing the voltage of phase R from 10% to 190%. The patterns are then formed in voltage-voltage planes. As the change of voltage has occurred in phase R, patterns developed by phase R and phase Y voltages and phase-B and phase-R voltages will be changed. Voltage signals and patterns generated in combination of R-Y and B-R phase voltages are formed; as the pattern of Y-B voltages remain unaltered, dimension of this pattern (Y-B) will not change. $[x_V]$ and $[y_V]$ for different voltage level in phase R are presented in Table 9.3.

Table 9.3 shows that unbalance due to phase R changes one element of [x] and one element of [y] and the element is not unity. From Table 9.3 an inference is drawn as a rule for a system having unbalance due to phase R voltage, which rule is presented in Table 9.4.

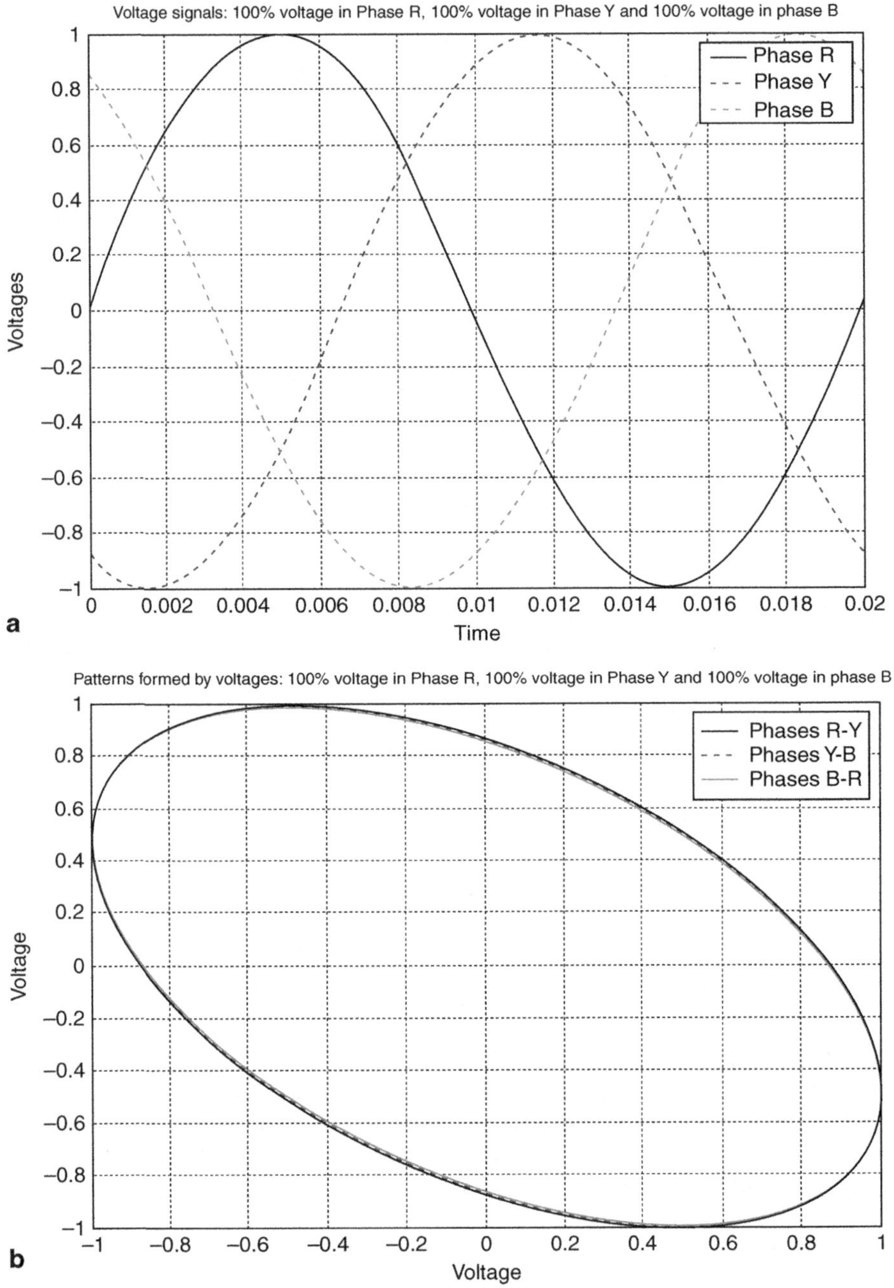

Fig. 9.2 **a** Voltage signals of a balanced system having normalized amplitudes (1:1:1). **b** Pattern developed by phase R voltage and phase Y voltage signals of Fig. 9.2a in voltage-voltage plane

Table 9.1 Values of $[x_V]$ and $[y_V]$ for a balanced system

Condition	Phase	$[y_V] = \begin{bmatrix} Y_{V_{RY}} \\ Y_{V_{YB}} \\ Y_{V_{BR}} \end{bmatrix}$	$[x_V] = \begin{bmatrix} X_{V_{RY}} \\ X_{V_{YB}} \\ X_{V_{BR}} \end{bmatrix}$
Balanced system	RY	1	1
	YB	1	1
	BR	1	1

Table 9.2 Rule developed from Table 9.1

Rule 1	$[x_V] = [111]$ and $[y_V] = [111]$	Power system is perfect balanced

Table 9.3 Values of $[x_V]$ and $[y_V]$ for unbalances in R phase

Condition	Phase	$[y_V]$	$[x_V]$
10% voltage in R phase	RY	1.0	0.1
	YB	1.0	1.0
	BR	0.1	1.0
30% voltage in R phase	RY	1.0	0.3
	YB	1.0	1.0
	BR	0.3	1.0
50% voltage in R phase	RY	1.0	0.5
	YB	1.0	1.0
	BR	0.5	1.0
70% voltage in R phase	RY	1.0	0.7
	YB	1.0	1.0
	BR	0.7	1.0
90% voltage in R phase	RY	1.0	0.9
	YB	1.0	1.0
	BR	0.9	1.0
110% voltage in R phase	RY	1.0	1.1
	YB	1.0	1.0
	BR	1.1	1.0
130% voltage in R phase	RY	1.0	1.3
	YB	1.0	1.0
	BR	1.3	1.0
150% voltage in R phase	RY	1.0	1.5
	YB	1.0	1.0
	BR	1.5	1.0
170% voltage in R phase	RY	1.0	1.7
	YB	1.0	1.0
	BR	1.7	1.0
190% voltage in R phase	RY	1.0	1.9
	YB	1.0	1.0
	BR	1.9	1.0

Table 9.4 Rule developed from Table 9.3

Rule 2	$[x_V] \neq [111]$ and $[y_V] \neq [111]$	Power system is unbalanced
Rule 3	If Rule 2 is true and $X_{RY} \neq 1$ and $Y_{BR} \neq 1$	Unbalance has occurred in phase R and percentage of unbalance $= (1 - X_{RY}) \times 100\%$

Case III: Unbalance in Phase Y Let, unbalance is introduced in Y phase voltage by gradually increasing voltage of phase Y from 10% to 190%. The patterns are formed in voltage-voltage planes. As the unbalance has occurred in phase Y, patterns developed by phase Y and phase B voltages and phase-R and phase-Y voltages will be changed. Voltage signals and patterns generated in combination of Y and B and R and Y phase voltages are formed. As the phase B and R voltages remain the same, dimension of patterns developed by phases B and R voltages will not change. $[x_V]$ and $[y_V]$ for different voltage levels of phase Y are presented in Table 9.5.

Table 9.5 shows that unbalance due to phase Y changes one element of $[x_V]$ and one element of $[y_V]$ and the element is not unity. From Table 9.5 an inference is drawn as a rule for a system having unbalance due to phase Y, which is presented in Table 9.6.

Table 9.5 Values of $[x_V]$ and $[y_V]$ for unbalances in Y phase

Condition	Phase	$[y_V]$	$[x_V]$
10% voltage in Y phase	RY	0.1	1.0
	YB	1.0	0.1
	BR	1.0	1.0
30% voltage in Y phase	RY	0.3	1.0
	YB	1.0	0.3
	BR	1.0	1.0
50% voltage in Y phase	RY	0.5	1.0
	YB	1.0	0.5
	BR	1.0	1.0
70% voltage in Y phase	RY	0.7	1.0
	YB	1.0	0.7
	BR	1.0	1.0
90% voltage in Y phase	RY	0.9	1.0
	YB	1.0	0.9
	BR	1.0	1.0
110% voltage in Y phase	RY	1.1	1.0
	YB	1.0	1.1
	BR	1.0	1.0
130% voltage in Y phase	RY	1.3	1.0
	YB	1.0	1.3
	BR	1.0	1.0
150% voltage in Y phase	RY	1.5	1.0
	YB	1.0	1.5
	BR	1.0	1.0
170% voltage in Y phase	RY	1.7	1.0
	YB	1.0	1.7
	BR	1.0	1.0
190% voltage in Y phase	RY	1.9	1.0
	YB	1.0	1.9
	BR	1.0	1.0

Table 9.6 Rule developed from Table 9.5

Rule 2	$[x_V] \neq [111]$ and $[y_V] \neq [111]$	Power system is unbalanced
Rule 4	If Rule 2 is true and $X_{YB} \neq 1$ and $Y_{RY} \neq 1$	Unbalance has occurred in phase Y and percentage of unbalance $= (1 - X_{YB}) \times 100\%$

Case IV: Unbalance in Phase B Let, unbalanced is now introduced in phase B by gradually increasing voltage of phase B from 10% to 190%. The patterns are formed, $[x_V]$ and $[y_V]$ for different voltage levels of phase B of those patterns are presented in Table 9.7.

Table 9.7 shows that unbalance due to phase B changes one element of $[x_V]$ and one element of $[y_V]$ and the element is not unity. From Table 9.7 an inference is

Table 9.7 Values of $[x_V]$ and $[y_V]$ for unbalances in B phase

Condition	Phase	$[y_V]$	$[x_V]$
10% voltage in B phase	RY	1.0	1.0
	YB	0.1	1.0
	BR	1.0	0.1
30% voltage in B phase	RY	1.0	1.0
	YB	0.3	1.0
	BR	1.0	0.3
50% voltage in B phase	RY	1.0	1.0
	YB	0.5	1.0
	BR	1.0	0.5
70% voltage in B phase	RY	1.0	1.0
	YB	0.7	1.0
	BR	1.0	0.7
90% voltage in B phase	RY	1.0	1.0
	YB	0.9	1.0
	BR	1.0	0.9
110% voltage in B phase	RY	1.0	1.0
	YB	1.1	1.0
	BR	1.0	1.1
130% voltage in B phase	RY	1.0	1.0
	YB	1.3	1.0
	BR	1.0	1.3
150% voltage in B phase	RY	1.0	1.0
	YB	1.5	1.0
	BR	1.0	1.5
170% voltage in B phase	RY	1.0	1.0
	YB	1.7	1.0
	BR	1.0	1.7
190% voltage in B phase	RY	1.0	1.0
	YB	1.9	1.0
	BR	1.0	1.9

Table 9.8 Rule developed from Table 9.7

Rule 2	$[x_V] \neq [111]$ and $[y_V] \neq [111]$	Power system is unbalanced
Rule 5	If Rule 2 is true and $X_{BR} \neq 1$ and $Y_{YB} \neq 1$	Unbalance has occurred in phase Y and percentage of unbalance $= (1 - X_{BR}) \times 100\%$

Table 9.9 Values of $[x_V]$ and $[y_V]$ for unbalances in R, Y and B phases combined

Condition	Phase	$[y_V]$	$[x_V]$
10% voltage in R and 50% voltage in Y phases	RY	0.5	0.1
	YB	1.0	0.5
	BR	0.1	1.0
10% voltage in R, 50% voltage in Y and 150% voltage in B phases	RY	0.5	0.1
	YB	1.5	0.5
	BR	0.1	1.5
30% voltage in R, 30% voltage in Y and 30% voltage in B phases	RY	0.3	0.3
	YB	0.3	0.3
	BR	0.3	0.3
40% voltage in R and 60% voltage in B phases	RY	1.0	0.4
	YB	0.6	1.0
	BR	0.4	0.6

drawn as a rule for a system having unbalance due to phase Y, which is presented in Table 9.8.

Case V: Unbalance in More Than One Phase For unbalance is introduced in more than one phases and patterns formed and $[x_V]$ and $[y_V]$ for different unbalanced signals presented in Table 9.9.

Table 9.10 Rules developed from Table 9.9

Rule nos.	Rule	Inference
Rule 2	$[x_V] \neq [111]$ and $[y_V] \neq [111]$	Power system is unbalanced
Rule 6	If Rule 2 is true and $X_{RY} \neq 1$, $X_{YB} \neq 1$ $Y_{RY} \neq 1$ and $Y_{BR} \neq 1$	Unbalance has occurred in phases R and Y, the percentage of unbalance in R phase $= (1 - X_{RY}) \times 100\%$ and the percentage of unbalance in Y phase $= (1 - X_{YB}) \times 100\%$
Rule 7	If Rule 2 is true and $X_{YB} \neq 1$, $X_{BR} \neq 1$, $Y_{YB} \neq 1$ and $Y_{RY} \neq 1$	Unbalance has occurred in phases Y and B, the percentage of unbalance in Y phase $= (1 - X_{YB}) \times 100\%$ and the percentage of unbalance in B phase $= (1 - X_{BR}) \times 100\%$
Rule 8	If Rule 2 is true and $X_{BR} \neq 1$, $X_{RY} \neq 1$, $Y_{YB} \neq 1$, $Y_{BR} \neq 1$	Unbalance has occurred in phases B and R, the percentage of unbalance in B phase $= (1 - X_{BR}) \times 100\%$ and the percentage of unbalance in R phase $= (1 - X_{RY}) \times 100\%$
Rule 9	If Rule 2 is true and $X_{RY} \neq 1$, $X_{YB} \neq 1$, $X_{BR} \neq 1$, $Y_{RY} \neq 1$, $Y_{YB} \neq 1$ and $Y_{BR} \neq 1$	Unbalance has occurred in phases R, Y, and B the percentage of unbalance in R phase $= X_{RY} \times 100\%$, the percentage of unbalance in Y phase $= (1 - X_{YB}) \times 100\%$ and the percentage of unbalance in B phase $= (1 - X_{BR}) \times 100\%$

Table 9.11 CMS rule set for unbalance assessment using FPEM

Rule nos.	Rule	Inference
Rule 1	$[x_V] = [111]$ and $[y_V] = [111]$	Power system is perfect balanced.
Rule 2	$[x_V] \neq [111]$ and $[y_V] \neq [111]$	Power system is unbalanced
Rule 3	If Rule 2 is true and $X_{RY} \neq 1$ and $Y_{BR} \neq 1$	Unbalance has occurred in phase R and percentage of unbalance $= (1 - X_{RY}) \times 100\%$
Rule 4	If Rule 2 is true and $X_{YB} \neq 1$ and $Y_{RY} \neq 1$	Unbalance has occurred in phase Y and percentage of unbalance $= (1 - X_{YB}) \times 100\%$
Rule 5	If Rule 2 is true and $X_{BR} \neq 1$ and $Y_{YB} \neq 1$	Unbalance has occurred in phase B and percentage of unbalance $= (1 - X_{BR}) \times 100\%$
Rule 6	If Rule 2 is true and $X_{RY} \neq 1$, $X_{YB} \neq 1$ $Y_{RY} \neq 1$ and $Y_{BR} \neq 1$	Unbalance has occurred in phases R and Y, the percentage of unbalance in R phase $= (1 - X_{RY}) \times 100\%$ and the percentage of unbalance in Y phase $= (1 - X_{YB}) \times 100\%$
Rule 7	If Rule 2 is true and $X_{YB} \neq 1$, $X_{BR} \neq 1$, $Y_{YB} \neq 1$ and $Y_{RY} \neq 1$	Unbalance has occurred in phases Y and B, the percentage of unbalance in Y phase $= (1 - X_{YB}) \times 100\%$ and the percentage of unbalance in B phase $= (1 - X_{BR}) \times 100\%$
Rule 8	If Rule 2 is true and $X_{BR} \neq 1$, $X_{RY} \neq 1$, $Y_{YB} \neq 1$, $Y_{BR} \neq 1$	Unbalance has occurred in phases B and R, the percentage of unbalance in B phase $= (1 - X_{BR}) \times 100\%$ and the percentage of unbalance in R phase $= (1 - X_{RY}) \times 100\%$
Rule 9	If Rule 2 is true and $X_{RY} \neq 1$, $X_{YB} \neq 1$, $X_{BR} \neq 1$, $Y_{RY} \neq 1$, $Y_{YB} \neq 1$ and $Y_{BR} \neq 1$	Unbalance has occurred in phases R, Y, and B, the percentage of unbalance in R phase $= (1 - X_{RY}) \times 100\%$, the percentage of unbalance in Y phase $= (1 - X_{YB}) \times 100\%$ and the percentage of unbalance in B phase $= (1 - X_{BR})$ 100%

Table 9.9 shows the conditions of unbalance in more than one phases. Here more than one elements of $[x_V]$ and $[y_V]$ are not unity. From Table 9.9, inferences are drawn as rules for a system with unbalance in more than one phase, which are presented in Table 9.10.

All above rules are combined and called, in the names of the developers, **Chattopadhyay, Mitra and Sengupta as CMS Rule Set for Unbalance Assessment using FPEM**. The rule set is presented in Table 9.11 [1–5].

9.5 Algorithm for Unbalance Assessment

Based on the **CMS Rule Set for Unbalance Assessment using FPEM**, an algorithm has been developed for unbalance assessment in real power system as follows:

1. Step down power system voltage and current signals by potential transformer and current transformer respectively.
2. Sample and normalize these data.
3. Digitize these data and receive in a central processing unit (CPU) through a Data Acquisition System (DAS).
4. Take data of phase voltages and line currents.

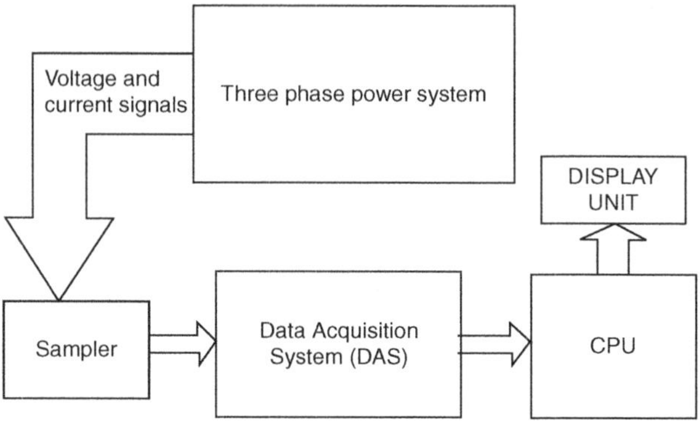

Fig. 9.3 Schematic diagram of experimental set up

5. Calculate $[x_V]$ and $[y_V]$ of the patterns developed in voltage-voltage and current-current planes.
6. Apply rule set and assess unbalance in voltage and current waveforms.

The schematic diagram of the experimental setup is shown in Fig. 9.3.

Power system data are reduced by potential transformer and current transformer. Data are collected through a data acquisition system (DAS). The data are captured by the central processing unit where they are processed and analyzed to develop feature pattern of combined phase voltages and currents in the display unit. $[x_V]$ and $[y_V]$ for voltage and current are calculated. Unbalance has been assessed with the help rule set of Table 9.11.

9.6 Discussion

In this chapter feature pattern extraction method has been used for unbalance assessment. Different phase voltages and currents are plotted in voltage-voltage planes. Dimension of patterns changes with the change of percentage of unbalance. Some (x and y) parameters have been defined. Two matrices [x] and [y] are formed for voltage. Changes of elements of these two matrices with the change of percentage of unbalance are noted. A rule set has been developed. The rule set has been applied for assessment of unbalance in real system. Use of rule set enables us to assess unbalance with the help of few parameters of [x] and [y]. This decreases the task of mathematical computation and save memory space required for storing data. Same analysis can be followed by considering the current signals drawn from real system.

References

[1] Chattopadhyay, S., Mitra, M., Sengupta, S.: Feature pattern extraction method for on line determination of percentage unbalance and the highest order of harmonic present in a three-phase system. Proceedings of An all India Seminar on Power System by IEEE, Calcutta section and IE, India (2003)

[2] Chattopadhyay, S., Mitra, M., Sengupta, S.: Computerized assessment of electric power quality. In Proceedings of 12th Science Congress organized by DST and Govt. of West Bengal in Kolkata, in February–March (2005)
[3] Chattopadhyay, S., Mitra, M., Sengupta, S.: IT based assessment of electric power quality of a power system. Proceedings of PEITSICON-2005: International Conference on IT and Power System, IEE (Calcutta section), Kolkata, January (2005)
[4] Chattopadhyay, S., Mitra, M., Sengupta, S.: Intelligent power quality monitoring in an electric power system. Best of 2004-volume on Modeling and Simulation, AMSE, France, pp. 18–22 (2004)
[5] Chattopadhyay, S., Mitra, M., Sengupta, S.: Power quality assessment in V-V, Clarke and Park domain. UPEC, pp. 669–676 (2007)

Chapter 10
Useful Tools for Harmonic Assessment

Abstract This chapter deals with the different mathematical tools used for harmonic assessment. In the introduction different aspect of harmonic assessment is discussed and classification of tools is done. Different aspects of Fourier transform are discussed. The limitation of this tool in the context of handling discrete signal is then discussed. To overcome this limitation Discrete Fourier Transform is discussed. For application purpose Fast Fourier Transform is mentioned. At last wavelet based transform technique is discussed.

10.1 Introduction

Harmonics is the most important issue among all issues related to power quality. Major sources and effects on power system have been described in the previous chapters. To analyze harmonics in power system, different new standards and definitions have been introduced; for example harmonic distortion factors of voltage and current waveform. However the basis of all harmonic assessment still depends on the measurement of amplitude and phase angle of the harmonics components. Order of harmonics present in the system may be integer or non integer multiple of the fundamental component. A lot of research is going on from the very past to assess both of these types of harmonics. Different mathematical tools have come out for this purpose. Some of them are capable of measuring both integer and non integer order of harmonics. Some of them are capable of measuring only integer type of harmonics. Also, in many cases, signals could not be captured in continuous form. To overcome these limitations modified mathematical tools have been developed to handle discrete signals. Mathematical tools [1] for harmonics analysis may be in time domain, or frequency domain or both time-frequency domain. Thus broadly, frequency assessment can be categories as follows

1. **Fourier Transform Based Assessment:** One of old techniques used in analysis of nonsinusoidal signals is Fourier transform. Fourier analysis has been used for power quality assessment for a long period. It permits mapping of signals

S. Chattopadhyay et al., *Electric Power Quality,* Power Systems,
DOI 10.1007/978-94-007-0635-4_10, © Springer Science+Business Media B.V. 2011

from time domain to frequency domain by decomposing the signals into several frequency components. Application of Discrete FT and Fast FT are very useful to overcome some of the disadvantages of the earlier one.

2. **Wavelet Transform Based Assessment:** Fourier transform fails in the analysis of transients owing to the non-stationary property of its signals in both time and frequency domains. Wavelet transform (WT) helps us in such cases. Wavelet analysis has been suggested as a new tool for measurement and monitoring power quality problems both in absence and presence of transients. Multi-resolution signal decomposition has been used to localize different power quality problems and assess them.
3. **Neuro-Fuzzy Based Assessment:** An ANN-fuzzy logic combined system for classifying power system disturbances are used to identify the event based quality issues. Fuzzy-Based Adaptive Digital Metering system and Genetic Algorithm, have been introduced to avoid effects of power quality problems.

10.2 Fourier Series

Baron Jean-Baptiste-Joseph Fourier (March 21, 1768–May 16, 1830) was born in a poor family in Auxerre in the era of Napoleon. Latter he became professor of mathematics. He used to believe that any arbitrary defined function could be expressed in a single analytical expression. Incidentally, the idea was introduced much before by Leonhard Euler (1707–1783). In this relation Euler published a formula

$$\frac{1}{2}x = \sin x - \frac{1}{2}\sin 2x + \frac{1}{3}\sin 3x \ + \cdots \tag{10.1}$$

But a very little credit goes to Euler. Fourier came with the idea and applied it in a real problem of heat transfer through solid material. In 1807, Fourier gave his theory of heat conduction and the expression for temperature distribution in solid medium. Unfortunately, during and after the publication, the theory faced a lot of doubts and criticism by Laplace, Lagrange, Biot and Poisson. However, slowly it was accepted by all and he got prize for this purpose in mathematics in 1811. Fourier also pointed out the limitation of Euler's expression and wrote, "The equation is no longer true when the value of x is between π and 2π. However, the second side of the equation is still a convergent series but the sum is not equal to x/2. Euler, who knew the equation and gave it without comment," (Quotation from J Herivel [2]).

Fourier expands continuous, single valued function of time domain in terms of dc component, and series of integer harmonics in 1822 [3–5]. As, power system voltage and current signals often consist of harmonics as well as dc components, postulate given by Fourier is frequently used to express them in terms of Fourier series. If f(t) be a time domain continuous periodic function, then it can be expressed as

$$f(t) = a_0 + \sum_{n=1,2,3,\ldots} a_n \cos n\omega t + \sum_{n=1,2,3,\ldots} b_n \sin n\omega t \tag{10.2}$$

represents the dc component present the function and n is the order of harmonics. The series is known as Fourier series. Coefficients a_0, a_n and b_n can be obtained as

$$a_0 = \frac{1}{T}\int_{-\frac{T}{2}}^{\frac{T}{2}} f(t)\,dt \tag{10.3}$$

$$a_n = \frac{2}{T}\int_{-\frac{T}{2}}^{\frac{T}{2}} f(t)\cos n\omega t\,dt \tag{10.4}$$

$$b_n = \frac{2}{T}\int_{-\frac{T}{2}}^{\frac{T}{2}} f(t)\sin n\omega t\,dt \tag{10.5}$$

For functions having odd symmetry,

$$b_n = \frac{4}{T}\int_{0}^{\frac{T}{2}} f(t)\sin n\omega t\,dt \tag{10.6}$$

For functions having even symmetry,

$$a_0 = \frac{2}{T}\int_{0}^{\frac{T}{2}} f(t)\,dt \tag{10.7}$$

$$a_n = \frac{4}{T}\int_{0}^{\frac{T}{2}} f(t)\cos n\omega t\,dt \tag{10.8}$$

In complex form f(t) can be expressed as

$$f(t) = \sum_{n=1,2,3\ldots}^{\propto} c_n e^{jn\omega t} \tag{10.9}$$

where, the coefficients are given by

$$c_n = \frac{1}{2}(a_n - jb_n) = \frac{1}{\pi}\int_{-\pi}^{\pi} f(\omega t)e^{jn\omega t}\,d(\omega t) \tag{10.10}$$

$$c_0 = a_0 - \frac{1}{2\pi}\int_{-\pi}^{\pi} f(\omega t)\,d(\omega t) \tag{10.11}$$

10.3 Fourier Transform

Time domain single valued, periodic and continuous function f(t) can be transformed into frequency domain by Fourier transformation method [3, 5].

$$F(f) = \int_{-\propto}^{+\propto} f(t)e^{-j2\pi fT}\,dt \tag{10.12}$$

provided

1. The function f(t) is single valued and continuous in well defined time interval T.
2. There is no discontinuity in time T, if there is any discontinuity, then it will be finite.
3. Number of maxima in time T is finite.

By Fourier transform, time domain function can be expressed in frequency domain the method is widely used to handle continuous signal in frequency domain. However, the transform method suffers from limitation to handle discrete or discontinuous, multi-valued and undefined signals which are often faced by electrical applications.

10.4 Discrete Fourier Transform

Most of the measurements of electrical parameters face signals in discrete forms which could not be analyzed by Fourier transform. Signals are available in the form of series of independent numerical values at discrete intervals. To deals with such discrete signals, Fourier transform (FT) is modified as discrete Fourier transforms (DFT). Frequency domain spectrums of a time domain function f(t) obtained by DFT is given as

$$F(f_k) = \frac{1}{N} \sum_{n=0}^{N-1} f(t_n) e^{j2\pi kn/N} \tag{10.13}$$

N is the total number of samples in one period. Here both time domain function and frequency domain spectrum are periodic. The frequency domain spectrums depend on the nature of sampling by which signals are being captured for measurement and analysis.

However, the DFT suffers from the main limitation of high execution time for large values of N.

10.5 Fast Fourier Transform

The limitation of DFT is overcome by Fast Fourier transform (FFT) [3]. By FFT, the matrix for multiplication required in DFT is resolved into many more simpler forms. This reduces the multiplication process as well as the time of execution. This finds a wide application area of FFT in measurement and analysis of frequency domain spectrum, specially, for electric signals which are captured in discrete form. For example by FFT, a 8-element DFT can be reduced to 2 four elements DFTs and further reduced to 4 two elements DFTs. However there will be still some error in DFT for the truncation of the data. There are many sophisticated mathematical software tools available in MATLAB by using FFT which can directly be applicable or applicable after some slight alteration on captured discrete signals. Fourier transform is then upgraded to short time Fourier transform (STFT) which deals with the wave dividing the time span into many short periods.

10.6 Hartley Transform and Discrete Hartley Transform

Like Fourier transform there is Hartley transform by which spectrums can be obtained in frequency domain [1]. The frequency domain function of a time function f(t) is given as

$$H(\omega) = \frac{1}{2\pi}\int_{-\alpha}^{\alpha} f(t)(\cos\omega t + \sin\omega t)\, dt \tag{10.14}$$

However, this function is related to the frequency domain function F(f) by the following relation

$$H(\omega) = Re[F(f)] - Im[F(f)] \tag{10.15}$$

Thus Hartley frequency function is the sum of real part of Fourier frequency function and sign inverse imaginary part of Fourier frequency function.

Inverse Hartley transform is given as

$$f(t) = \frac{1}{2\pi}\int_{-\alpha}^{\alpha} H(\omega)(\cos\omega t + \sin\omega t)\, d\omega \tag{10.16}$$

Like DFT, discrete Hartley transform (DHT) is given as

$$H(f_k) = \frac{1}{N}\sum_{n=0}^{N-1} f(t_n)\cos 2\pi kn/N \tag{10.17}$$

The factor of DHT is similar to the DFT. Apart from this factor, DHT is real and symmetrical. As further advancement of DHT, complex DHT, Cas-Cas transform (CCT) and then discrete sine and cosine Transform (DSCT) were introduced.

10.7 Wavelet Transform

Wavelet transform (WT) has now been well accepted tool for analyzing signals [1, 6]. It has the following advantages:

- WT can deal with non-stationary waves.
- WT is fast.
- WT can decompose a signal into its frequency components.
- WT can tailor frequency resolution which is helpful to identify sources of transients.

- WT can focus on low interval high frequency components and long interval low frequency components.
- WT can focus on localized impulses and oscillations.

In WT, there is mother functions g(t) which can be altered and applied in the analysis. Thus, WT application is based on the mother function g(t). The function g(t) is a combination of a decaying function and an oscillatory function, mathematically it can be expressed as

$$g(t) = e^{-\alpha t^2} e^{j\omega t} \tag{10.18}$$

Let us derive a function from the mother function in the following way,

$$g'(a, b, t) = \frac{1}{\sqrt{a}} g\left(\frac{t-b}{a}\right) \tag{10.19}$$

Based on this derived function WT of a function f(t) can be written as

$$WT(a, b) = \frac{1}{\sqrt{a}} \int_{-\infty}^{\infty} f(t)\, g\left(\frac{t-b}{a}\right) dt \tag{10.20}$$

In wavelet analysis de-noising techniques are very important. To deal with discrete data, wavelet transform is then upgraded to discrete wavelet transform (DWT). For analysis with the help of some other real or predefined signal, wavelet transform is modified as cross wavelet transform which deals two wave simultaneously.

10.8 Discussion

Various mathematical tools used for analysis of lower and higher frequency have been discussed. DFT is advantageous as it can deal with discrete signals. Execution speed of DFT is increased in FFT. For this reason, among them FFT is now extensively used in spectrum analyzer. DHT is an alternate of DFT. However, use of WT has increased to a great extent which has some advantages over the others. WT can focus on localized frequency and oscillations.

References

[1] Jon, A., Neville, R.W.: Power System Harmonics, 2nd ed. Wiley (2003)
[2] Herivel, J.: Joseph Fourier: The Man and the Physicist, p. 319. Clarendon Press, Oxford (1975)
[3] Fourier, J.B.J.: Theorie analytique de la chaleur. Paris (1822)
[4] Brigham, E.O.: The Fast Fourier Transform. Prentice Hall, Englewood Cliffs (1974)
[5] Bracewell R.N.: The Fourier Transform and Its Application, 3rd ed. Mc Graw Hill, New York (2000)
[6] Eugenio, H., Guido, W.: A First Course on Wavelets. CRC Press, Boca Raton (1996)

Chapter 11
Harmonic Assessment Using FPEM in V-V and I-I Planes

Abstract This chapter describes an approach for analysis of harmonics present in a power system using Feature Pattern Extraction Method. The theoretical background for assessment of harmonics using this method in V-V and I-I planes is discussed. Computer simulation results are presented out of which rule sets are generated. Limitations present in the generated rule set are also highlighted. An algorithm has been developed for assessment of the highest order of harmonics followed by a general discussion.

11.1 Introduction

In this chapter, feature pattern extraction method is applied to assess harmonics present in voltage and current of a power system through feature pattern extraction method. In Chap. 9, it was shown that any two phase-voltages and line-currents form a closed loop in voltage-voltage and current-current plane. In those patterns harmonics were not considered and some elliptical patterns were developed by harmonic free voltages and currents. The rule set developed was able to assess unbalance. In this chapter, patterns are formed by voltages and currents consisting of fundamental component as well as harmonic components. Patterns show that cleavages appear in presence of harmonics. These cleavages carry specific information of harmonics present in the system. Features have been extracted, wherefrom a rule has been generated to find the highest order of dominating harmonics. An algorithm has also been developed to apply the rule sets in harmonic assessment in real power system. Limitations present in the rule have been discussed.

11.2 Harmonic Assessment by FPEM

In presence of harmonics, voltage and current waveforms of a power system may be written as

$$v(t) = \sum_{n=1,2,3\ldots n_V} V_n \sin(n\omega t - \varphi_n) \tag{11.1}$$

S. Chattopadhyay et al., *Electric Power Quality,* Power Systems,
DOI 10.1007/978-94-007-0635-4_11,

$$i(t) = \sum_{n=1,2,3\ldots n_I} I_n \sin(n\omega t - \theta_n) \tag{11.2}$$

where, n is order of harmonics, n_V is the highest order of harmonic present in voltage waveform, n_I is the highest order of harmonic present in current waveform, ω corresponds to fundamental frequency. In a three phase system, voltages are written as

$$\begin{bmatrix} v_R \\ v_Y \\ v_B \end{bmatrix} = \begin{bmatrix} \sum\limits_{m=1,2,3,\ldots\ldots} V_{Rm} \sin(m\omega t - \varphi_{Rm}) \\ \sum\limits_{m=1,2,3,\ldots\ldots} V_{Ym} \sin\left(m\omega t - \frac{2\pi}{3} - \varphi_{Ym}\right) \\ \sum\limits_{m=1,2,3,\ldots\ldots} V_{Bm} \sin\left(m\omega t - \frac{4\pi}{3} - \varphi_{Bm}\right) \end{bmatrix} \tag{11.3}$$

Similarly, currents in a three phase system are

$$\begin{bmatrix} i_R \\ i_Y \\ i_B \end{bmatrix} = \begin{bmatrix} \sum\limits_{n=1,2,3,\ldots\ldots} I_{Rn} \sin(n\omega t - \theta_{Rn}) \\ \sum\limits_{n=1,2,3,\ldots\ldots} I_{Yn} \sin\left(n\omega t - \frac{2\pi}{3} - \theta_{Yn}\right) \\ \sum\limits_{n=1,2,3,\ldots\ldots} I_{Bn} \sin\left(n\omega t - \frac{4\pi}{3} - \theta_{Bn}\right) \end{bmatrix}$$

Patterns have been formed with any two voltages and any two currents in voltage-voltage and current-current planes. The presence of harmonics brings cleavages in the patterns [1]. Nature of these cleavages depends on the amplitude and order of harmonics present in the system. The number of cleavages will change with the change of highest order of harmonic component of the system.

11.3 Patterns in V-V Planes in Presence of Harmonic

Voltage and current waveforms consisting of harmonics form elliptical patterns that have cleavages at the periphery. Some such patterns formed by normalized voltage signals having harmonics are shown in Fig. 11.1. Figures in (a) shows the signals and figures in (b) show the patterns form in voltage-voltage plane by those signals. These patterns show that the presence of harmonics brings cleavages in the closed patterns.

Nature of these cleavages depends on the amplitude and order of harmonics present in the system [2]. The number of cleavages is changing with the change of highest dominating order of harmonic component of the system. A definite relation between order of that harmonic component and number of cleavages has been found [3] and is presented in Table 11.1.

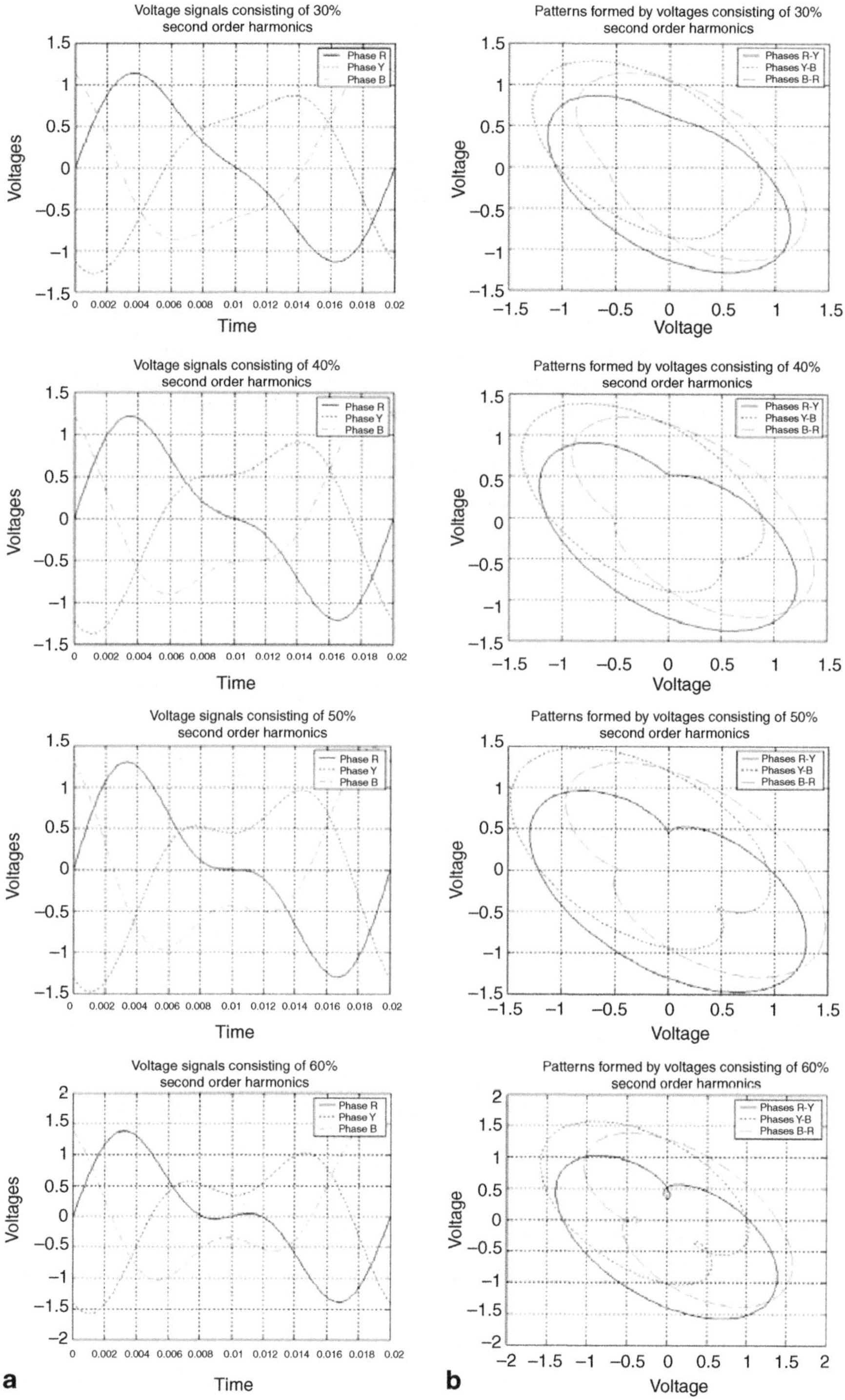

Fig. 11.1 **a** Normalized voltage signals and **b** patterns in V-V plane

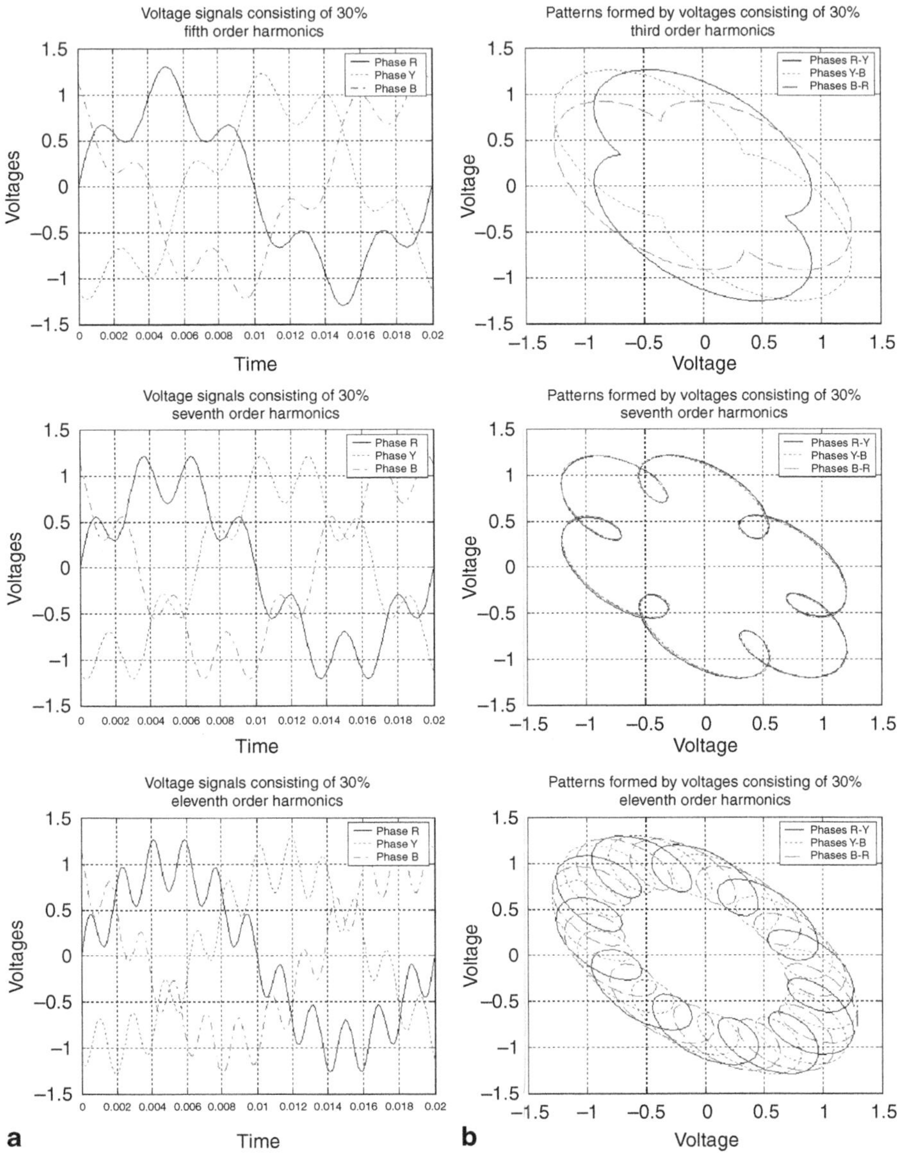

Fig. 11.1 (Continued)

11.4 CMS Rule for Determination of Highest order of Dominating Harmonics

The order of harmonic component and number of cleavages have been presented in Table 11.1, wherefrom, **CMS Rule for Determination of Highest order Dominating Harmonics** is formed. The rule is given in Table 11.2.

Table 11.1 Number of cleavages (C) linked with order of highest harmonic in a balanced system

Orders of harmonics present in the system	Number of cleavages (C)	Highest order of harmonic (n_H)
2	1	2
3	2	3
2, 3	2	3
2, 3,	2	3
2, 3, 5	4	5
3, 5	4	5
5, 7	6	7
5, 7, 13	12	13
2, 3, 5, 6, 7, 13	12	13

Table 11.2 CMS rule for determination of highest order dominating harmonics

C = number of cleavages in a pattern	Order of highest harmonic ($n_H = C + 1$)

It is also observed that the angular position of the cleavages is changing with amplitude and order of harmonics. Inter distance between any two consecutive cleavages and depth of a cleavage depends on the amplitude of the harmonic components. But, unfortunately, no definite relation has been found to assess the actual amplitude from these cleavages of the patterns.

11.5 Limitation of FPEM for Harmonic Assessment in V-V and I-I Plane

The limitation of using this method for harmonic assessment from features drawn in voltage-voltage and current-current planes with phase voltages and currents is it's inability to measure the amplitude of harmonic components present in the system from the patterns developed in these planes. Thus, assessment of harmonic distortion, harmonic power, etc, is not possible by FPEM in V-V or I-I planes.

11.6 Algorithm for Real Power System Data [1]

1. Step down power system voltage and current signals by potential transformer and current transformer respectively.
2. Sampled and normalize these data.
3. Digitize these data and receive in a central processing unit (CPU) through a Data Acquisition System (DAS).
4. Take data of phase voltages and line current.

5. Take data of any two phase-voltages and plot in voltage-voltage plane for one complete cycle.
6. Similarly, take data of any two line-currents and plot in current-current plane for one complete cycle.
7. Observe whether any cleavage(s) appears in the patterns developed in voltage-voltage and current-current planes.
8. If there is any cleavage, then it can be concluded that the system consists of harmonics. Otherwise the system is free from harmonics.
9. If there is harmonics in the system then, calculate highest order of harmonics using rule of Table 11.2.

11.7 Discussions

The patterns developed in this chapter may be formed in voltage-voltage and current-current planes. Effects of harmonics in the patterns have been observed. A rule has been set to measure order of highest harmonic present in the system. The change in patterns with respect to change in amplitude of harmonics is very complex and it does not generate any rule for assessment of percentage of harmonics present in the system. Thus, the above approach is only capable of giving a clear indication whether there is any harmonic in the system. If there is any harmonic in the system, the algorithm is also capable to measure the highest order of harmonics.

References

[1] Chattopadhyay, S., Mitra, M., Sengupta, S.: IT based assessment of electric power quality of a power system. Proceedings of PEITSICON-2005: International Conference on IT and Power System, IEE (Calcutta Branch), Kolkata, Paper No. 83. January 28, 29, 2005
[2] Chattopadhyay, S., Mitra, M., Sengupta, S.: Intelligent Power Quality Monitoring in an Electric Power System, Vol.—Best of 2004, on Modeling and Simulation, AMSE (France), pp. 146–156 (2004)
[3] Chattopadhyay, S., Mitra, M., Sengupta, S.: Feature pattern extraction method for on line determination of percentage unbalance and the highest order of harmonic present in a three-phase system. Proceedings of an All India Seminar on Integrated Operation of Power System, IEEE (Calcutta section), PE Chapter and IE (India), pp. 34–37, September 6–9, 2003

Chapter 12
Clarke and Park Transform

Abstract This chapter presents a brief idea of Clarke and Park transformations in which phase currents and voltages are expressed in terms of current and voltage space vectors. The space vectors are then represented in stationary reference frame. Then general rotating frame of reference has been introduced. The rotating frame of reference is then described in terms of d and q axes. The space vector is then expressed with respect to d-q reference frame. Clarke, Park and Inverse Park transformations have been described. These transformations are used in the subsequent chapters for assessment of power quality items.

12.1 Introduction

Clarke and Park transformations are used in high performance architectures in three phase power system analysis. Current and voltage are represented in terms of space vector which is represented in a stationary reference frame. A general rotating reference frame has then been introduced. This frame is described by d and q axes Clarke, Park and Inverse Park transformations have been described [1, 2]. Through the use of the Clarke transformation, the real and imaginary currents can be identified. The Park transformation is used to realize the transformation of those real and imaginary currents from the stationary to the rotating reference frame.

12.2 Current Space Vector

Let, i_R, i_Y and i_B are the instantaneous balanced three-phase currents. Then,

$$i_R + i_Y + i_B = 0 \tag{12.1}$$

Current space vector can be represented in terms of phase currents as

$$\bar{i} = k(i_R + a i_Y + a^2 i_B) \tag{12.2}$$

S. Chattopadhyay et al., *Electric Power Quality*, Power Systems,
DOI 10.1007/978-94-007-0635-4_12,

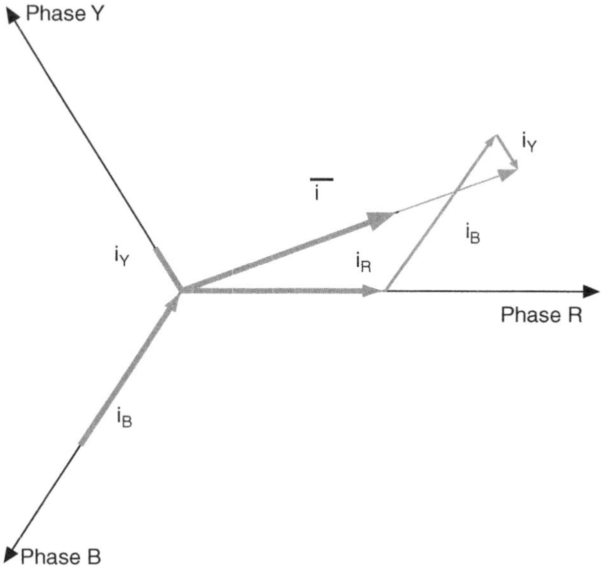

Fig. 12.1 Current space vector and its projection

where, a is an operator described earlier and k = Transformation constant. Figure 12.1 shows space current vector and its projection.

12.3 Stationary Reference Frame

The space vector defined by (12.2) can be expressed utilizing two-axis theory. The real part of the space vector is equal to the instantaneous value of the direct-axis current component, i_α, and imaginary part is equal to the quadrature-axis current component, i_β. This is shown in Fig. 12.2.

Thus, the current space vector, in the stationary reference frame can be expressed as:

$$\bar{i} = (i_\alpha + i_\beta) \tag{12.3}$$

In symmetrical 3-phase machines, the direct and quadrature axis currents i_α and i_β are fictitious quadrature-phase (2-phase) current components, which are related to the actual 3-phase currents as

$$i_\alpha = k\left(i_R - \frac{1}{2}i_Y - \frac{1}{2}i_B\right) \tag{12.4}$$

$$i_\beta = k\frac{\sqrt{3}}{2}(i_Y - i_B) \tag{12.5}$$

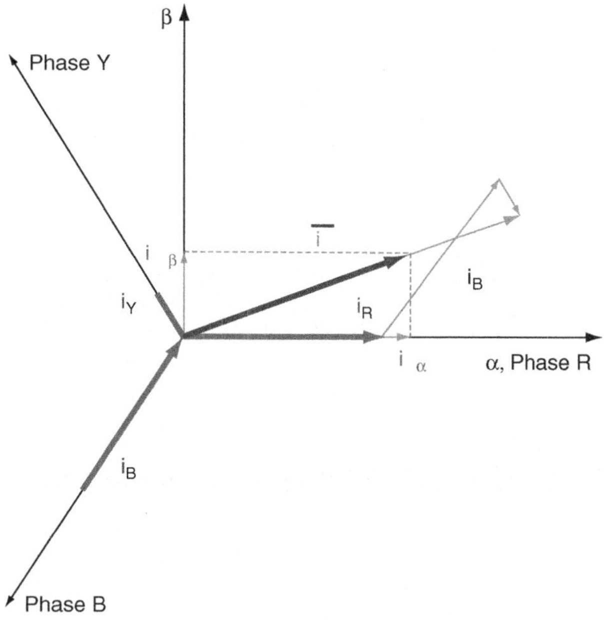

Fig. 12.2 Current space vector in (α, β) plane

In matrix form, the stator current in the stationary reference frame in terms of three phase currents can be written as

$$\begin{bmatrix} i_\alpha \\ i_\beta \end{bmatrix} = k \times \begin{bmatrix} 1 & -\frac{1}{2} & -\frac{1}{2} \\ 0 & \frac{\sqrt{3}}{2} & -\frac{\sqrt{3}}{2} \end{bmatrix} \times \begin{bmatrix} i_R \\ i_Y \\ i_B \end{bmatrix} \tag{12.6}$$

If the three-phase system is symmetrical

$$i_R + i_Y + i_B = 0 \tag{12.7}$$

Then,

$$i_\alpha \alpha = k\left(i_R - \frac{1}{2} i_Y - \frac{1}{2} i_B\right) \tag{12.8}$$

The recommended value of k is

$$k = \frac{2}{3} \tag{12.9}$$

Thus, transformations from a 3-phase (R, Y, B) to a 2-phase (α, β) system is commonly known as Clarke transformation. Thus, (12.6) can be written as

$$\begin{bmatrix} i_\alpha \\ i_\beta \end{bmatrix} = \frac{2}{3} \times \begin{bmatrix} 1 & -\frac{1}{2} & -\frac{1}{2} \\ 0 & \frac{\sqrt{3}}{2} & -\frac{\sqrt{3}}{2} \end{bmatrix} \times \begin{bmatrix} i_R \\ i_Y \\ i_B \end{bmatrix} \tag{12.10}$$

The transformation matrix is known as Clarke matrix or Clarke transformation matrix and the plane (α, β) is known as Clarke plane. So,

$$\begin{bmatrix} i_\alpha \\ i_\beta \end{bmatrix} = [Clarke\ Matrix] \times \begin{bmatrix} i_R \\ i_Y \\ i_B \end{bmatrix} \tag{12.11}$$

where,

$$[Clarke\ Matrix] = \frac{2}{3} \times \begin{bmatrix} 1 & -\frac{1}{2} & -\frac{1}{2} \\ 0 & \frac{\sqrt{3}}{2} & -\frac{\sqrt{3}}{2} \end{bmatrix} \tag{12.12}$$

12.4 General Rotating Reference Frame

Besides the stationary reference frame attached to the stator, current space vector equations can be formulated in a general reference frame which rotates at a general speed ω_G as shown in Fig. 12.3.

If a general reference frame is used, with direct and quadrature axes (*x and y*) rotating at a general instantaneous speed, $\omega_G = d\theta_G/dt$, as shown in Fig. 12.3, where θ_G is the angle between the direct axis of the stationary reference frame (α) attached to the real axis (x) of the general reference frame, then, the current space vector in general reference frame can be written as

$$\bar{i}_G = \bar{i}e^{-j\theta_G} = (i_x + i_y) \tag{12.13}$$

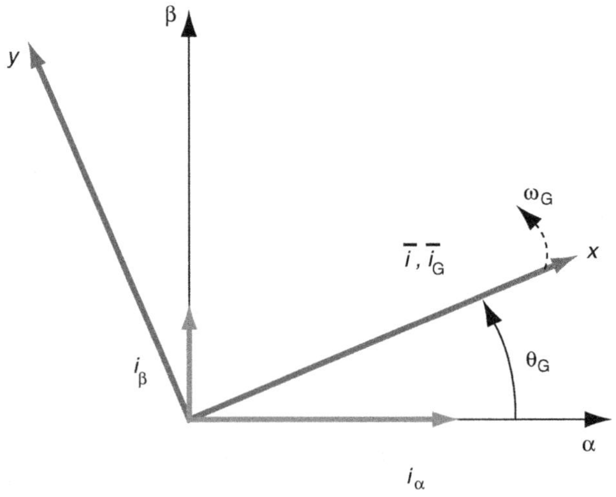

Fig. 12.3 General rotating frame of reference

12.5 d-q Rotating Reference Frame

Let us now convert x and y axes into d and q axes. Assume ψ be a vector along d-axis as shown in Fig. 12.4

In (α, β) plane,

$$\psi = \psi_\alpha + \psi_\beta \tag{12.14}$$

In (d, q) plane,

$$\psi = \psi_d + \psi_q \tag{12.15}$$

Angle between (α, β) and (d, q) is θ. Then

$$\sin\theta = \frac{\psi_\beta}{\psi_d} \tag{12.16}$$

$$\cos\theta = \frac{\psi_\alpha}{\psi_d} \tag{12.17}$$

The following transformations are involved due to rotation of orthogonal d-q system

1. α-β to d-q: Park transformation
2. d-q to α-β: Inverse Park transformation

Transformation from (α, β) to (d, q) is done by

$$\begin{bmatrix} d \\ q \end{bmatrix} = \begin{bmatrix} \cos\theta & \sin\theta \\ -\sin\theta & \cos\theta \end{bmatrix} \times \begin{bmatrix} \alpha \\ \beta \end{bmatrix} \tag{12.18}$$

or,

$$\begin{bmatrix} d \\ q \end{bmatrix} = [Park\ Matrix] \times \begin{bmatrix} \alpha \\ \beta \end{bmatrix} \tag{12.19}$$

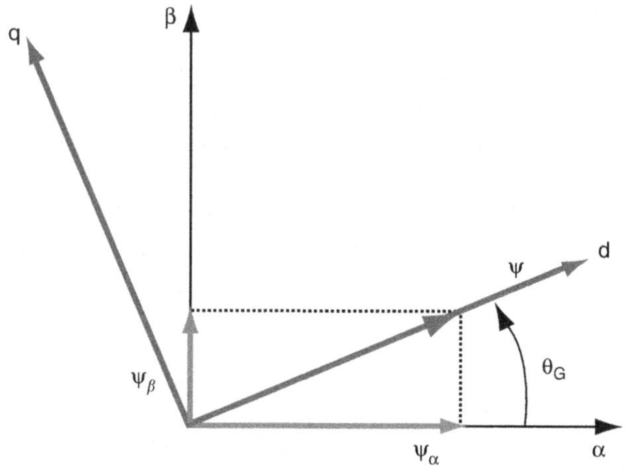

Fig. 12.4 d-q rotating frame of reference

where,

$$[Park\ Matrix] = \begin{bmatrix} \cos\theta & \sin\theta \\ -\sin\theta & \cos\theta \end{bmatrix} \tag{12.20}$$

Transformation from (d, q) to (α, β) is done by

$$\begin{bmatrix} \alpha \\ \beta \end{bmatrix} = \begin{bmatrix} \cos\theta & -\sin\theta \\ \sin\theta & \cos\theta \end{bmatrix} \times \begin{bmatrix} d \\ q \end{bmatrix} \tag{12.21}$$

or,

$$\begin{bmatrix} \alpha \\ \beta \end{bmatrix} = [Inverse\ Park\ Matrix] \times \begin{bmatrix} d \\ q \end{bmatrix} \tag{12.22}$$

where,

$$[Inverse\ Park\ Matrix] = \begin{bmatrix} \cos\theta & -\sin\theta \\ \sin\theta & \cos\theta \end{bmatrix} \tag{12.23}$$

12.6 Transformation Matrices

The transformations described in previous sections can be summarized in general form as

$$\begin{bmatrix} \alpha \\ \beta \end{bmatrix} = [Clarke\ Matrix] \times \begin{bmatrix} R \\ Y \\ B \end{bmatrix} \tag{12.24}$$

$$\begin{bmatrix} d \\ q \end{bmatrix} = [Park\ Matrix] \times \begin{bmatrix} \alpha \\ \beta \end{bmatrix} \tag{12.25}$$

$$\begin{bmatrix} d \\ q \end{bmatrix} = [Park\ Matrix] \times [Clarke\ Matrix] \times \begin{bmatrix} R \\ Y \\ B \end{bmatrix} \tag{12.26}$$

The transformation between R-Y-B, Clarke plane and Park plane is shown by a block diagram in Fig. 12.5. These transformations can be applied both for phase currents and phase voltages. First consider phase currents with respect to all frames of reference as shown in Fig. 12.6.

Currents in Clarke plane can be obtained from phase currents as follows

$$\begin{bmatrix} i_\alpha \\ i_\beta \end{bmatrix} = [Clarke\ Matrix] \times \begin{bmatrix} i_R \\ i_Y \\ i_B \end{bmatrix} \tag{12.27}$$

Currents in Park plane can be obtained from Clarke plane currents as follows

$$\begin{bmatrix} i_d \\ i_q \end{bmatrix} = [Park\ Matrix] \times \begin{bmatrix} i_\alpha \\ i_\beta \end{bmatrix} \tag{12.28}$$

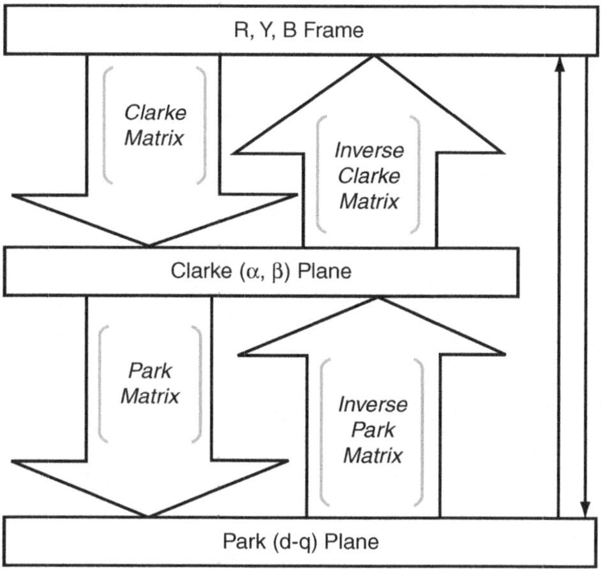

Fig. 12.5 Transformation of reference frame

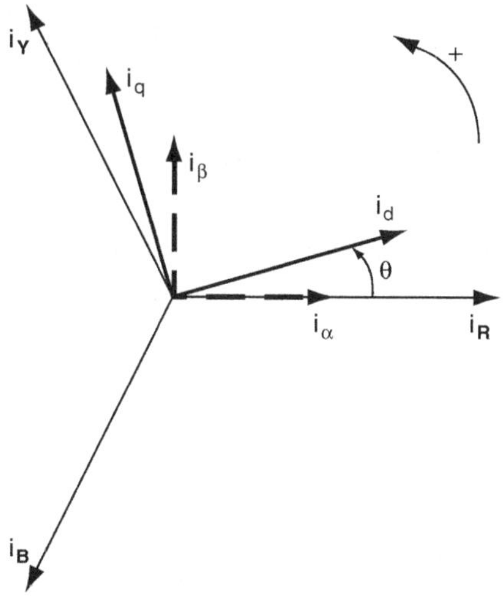

Fig. 12.6 R-Y-B, (α, β) and (d, q) current reference frames

Now, consider phase voltages with respect to all frames of reference as shown in Fig. 12.7.

Voltages in Clarke plane can be obtained from phase voltages as follows

$$\begin{bmatrix} v_\alpha \\ v_\beta \end{bmatrix} = [Clarke\ Matrix] \times \begin{bmatrix} v_R \\ v_Y \\ v_B \end{bmatrix} \tag{12.29}$$

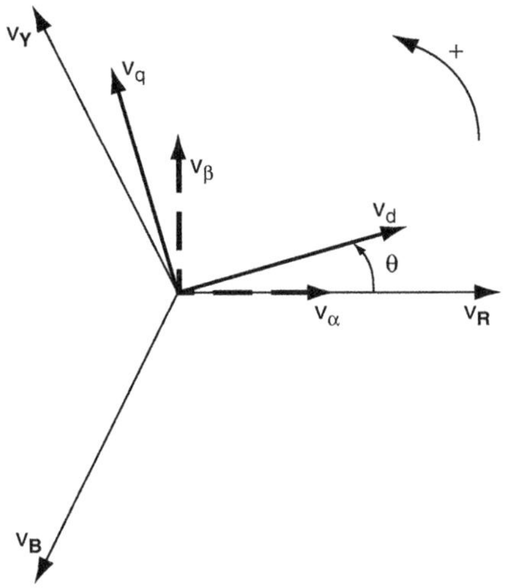

Fig. 12.7 R-Y-B, (α, β) and (d, q) voltage reference frames

Voltages in Park plane can be obtained from Clarke plane voltages as follows

$$\begin{bmatrix} v_d \\ v_q \end{bmatrix} = [Park\ Matrix] \times \begin{bmatrix} v_\alpha \\ v_\beta \end{bmatrix} \tag{12.30}$$

12.7 Discussion

From the above sections, it is seen that the voltage and current in a power system can be converted to a stationary reference frame and also to a rotating reference frame using Clarke transformation and Park transformation respectively. Original signals can be restored by inverse Clarke and inverse Park transforms. Applications of these transformations have specific advantages some of which are described in the next chapters.

References

[1] Ferrero A., Morando A. P., Ottoboni R., Superti-Furga G., Willems J. L.: On the meaning of the park power components in three-phase systems under non-sinusoidal conditions. Eur. Trans. Electr. Power Eng. **3**(1), 33–43 (1993)

[2] Akagi H., Nabae A.: The p-q theory in three-phase systems under non-sinusoidal conditions. Eur. Trans. Electr. Power Eng. **3**(1), 27–31 (1993)

Chapter 13
Harmonics Assessment by FPEM in Clarke and Park Planes

Abstract Starting with an introduction, this chapter describes the theory of analysis of harmonics present in power system through feature pattern extraction in Clarke plane. It then goes to the computer simulation part, wherefrom rule set is generated for assessment of harmonics present with or without phase shift of harmonics with reference to the fundamental. After this, it describes harmonic analysis in Park plane using same feature pattern extraction method. Theoretical developments are made first; computer simulation is then done to generate rule set.

13.1 Introduction

This chapter deals with some attempts to overcome the limitations of harmonic assessment in voltage-voltage and current-current plane using feature pattern extraction method. In Chap. 11, patterns are developed in voltage-voltage and current-current planes. In these patterns, cleavages appear in presence of harmonics. Number of cleavages is related to the order of highest harmonics. Depth and angular orientation of cleavages are related both to the amplitude and order of harmonics. But the nature of this relation is very complex and so, it is very difficult to form a rule set to assess amplitude of harmonic components. To overcome these limitations, in this chapter, phase-voltages and line-currents are transformed into Clarke and Park planes or domains. Patterns have been formed in Clarke and Park domains [1]. Patterns for harmonic free balanced system are circular in nature. As a result cleavages due to harmonics occur in particular fashion and at some specific angles [2, 3]. It is, therefore, easier to trace the behavior of cleavages. Thus, study of cleavages formed in patterns enables us better to get information of harmonics present in the system.

S. Chattopadhyay et al., *Electric Power Quality,* Power Systems,
DOI 10.1007/978-94-007-0635-4_13,

13.2 Harmonic Analysis in Clarke Plane

A good quality power consists of balanced voltage and current waveforms which are free from harmonics. Voltages can be represented as

$$v_R(t) = V_R \sin \omega t \tag{13.1}$$

$$v_Y(t) = V_Y \sin(\omega t + 120^\circ) \tag{13.2}$$

$$v_B(t) = V_B \sin(\omega t + 240^\circ) \tag{13.3}$$

Similarly, currents can be represented as

$$i_R(t) = I_R \sin(\omega t + \phi) \tag{13.4}$$

$$i_Y(t) = I_Y \sin(\omega t + 120^\circ + \phi) \tag{13.5}$$

$$i_B(t) = I_B \sin(\omega t + 240^\circ + \phi) \tag{13.6}$$

For a balanced system, $V_R = V_Y = V_B$ and $I_R = I_Y = I_B$

Voltage in Clarke Plane (α, β plane) can be found out by multiplying phase voltages by Clark transformation matrix as follows:

$$\begin{bmatrix} V_\alpha \\ V_\beta \end{bmatrix} = [Clarke\ Matrix] \times \begin{bmatrix} V_R \sin \omega t \\ V_Y \sin(\omega t - 2\pi/3) \\ V_B \sin(\omega t - 4\pi/3) \end{bmatrix} \tag{13.7}$$

For balanced system, $V_R = V_Y = V_B = V$ (say). Then, from (13.7)

$$\begin{bmatrix} V_\alpha \\ V_\beta \end{bmatrix} = V \times [Clarke\ Matrix] \times \begin{bmatrix} \sin \omega t \\ \sin(\omega t - 2\pi/3) \\ \sin(\omega t - 4\pi/3) \end{bmatrix} \tag{13.8}$$

For normalized waveforms, V = 1. Then, from (13.8), voltages in Clarke (α, β) plane are given as

$$\begin{bmatrix} V_\alpha \\ V_\beta \end{bmatrix} = [Clarke\ Matrix] \times \begin{bmatrix} \sin \omega t \\ \sin(\omega t - 2\pi/3) \\ \sin(\omega t - 4\pi/3) \end{bmatrix} \tag{13.9}$$

If any harmonic is present in each phase of the system, then (13.9) can be written as

$$\begin{bmatrix} V_\alpha \\ V_\beta \end{bmatrix} = [Clarke\ Matrix] \times \begin{bmatrix} \sin \omega t + X \sin n\omega t \\ \sin(\omega t - 2\pi/3) + X \sin(n\omega t - 2\pi/3) \\ \sin(\omega t - 4\pi/3) + X \sin(n\omega t - 4\pi/3) \end{bmatrix} \tag{13.10}$$

where, X is the magnitude of harmonic expressed as percentage of fundamental and n is the order of harmonic.

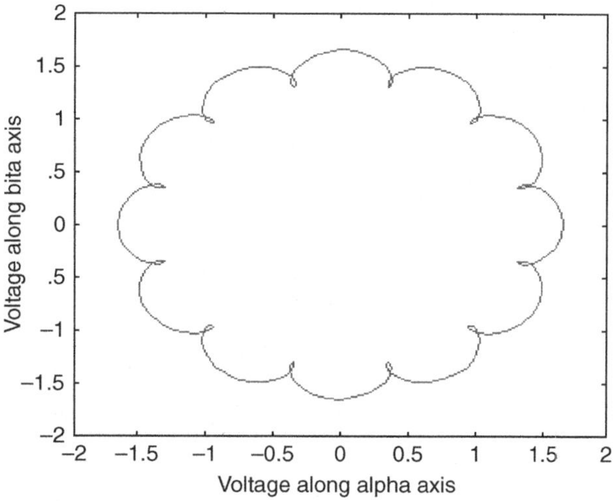

Fig. 13.1 Pattern formed in Clarke plane by system voltages in presence of harmonics

Due to the presence of the harmonics, voltages of (13.10) form a closed loop when plotted in Clarke (α, β) plane which will consist of cleavage(s) as shown in Fig. 13.1 [4].

The cleavages carry the information of harmonics present in voltage waveforms. If the harmonics have some phase difference with fundamental, then (13.10) becomes

$$\begin{bmatrix} V_\alpha \\ V_\beta \end{bmatrix} = \frac{2}{3}[Clarke\ Matrix] \times \begin{bmatrix} \sin \omega t + X \sin (n\omega t - \varphi_n) \\ \sin (\omega t - 2\pi/3) + X \sin (n\omega t - 2\pi/3 - \varphi_n) \\ \sin (\omega t - 4\pi/3) + X \sin (n\omega t - 4\pi/3 - \varphi_n) \end{bmatrix} \quad (13.11)$$

Variation of phase angles in the harmonic components will change the angular positions of the cleavages.

Algorithm The algorithm for harmonic assessment following algorithm is as follows

1. Take power system data from secondary sides of current and potential transformer.
2. Normalize voltage and current waveforms
3. Transform voltage and current waveforms into Clarke plane.
4. Plot data in Voltage-voltage and current-current planes in Clarke domain.
5. Generate feature patterns of different signals in Clarke plane.
6. Make observation for harmonic assessment.

a. Signals Having Harmonics with no Phase Shift from the Fundamental: With the increase of percentage of harmonics more patterns are formed as shown in Fig. 13.2. It shows that the number of cleavage remains same, but the depth of cleavage increases with the increase of percentage of harmonics.

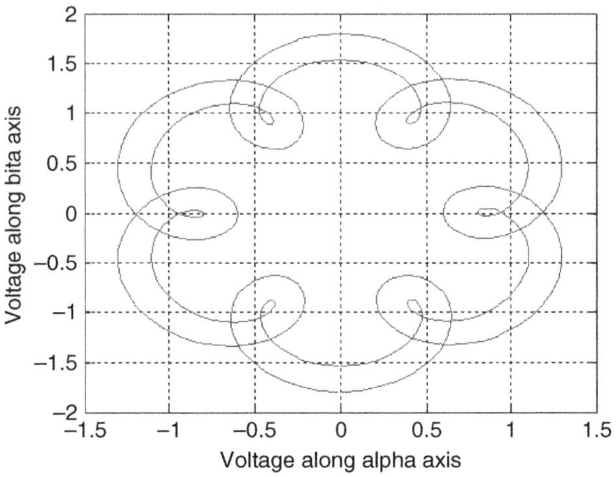

Fig. 13.2 Patterns formed by voltage signals having increasing amplitude of seventh order harmonics in Clarke plane

Harmonic brings cleavage(s) in the closed loop formed by V_α and V_β. To make a relation between order of harmonics and cleavage, following parameters are defined:

n = order of harmonic
C = Number of cleavage(s)
θ_C = Angular separation between two consecutive cleavage(s)

- Pattern formed by V_α and V_β of a system having second order harmonics shows that the pattern has one cleavage and it is located at 270°.
- Pattern formed by V_α and V_β of a system having third order harmonics shows that the pattern has two cleavages. Angular separation between two consecutive cleavage (θ_C) = 180°. One of the cleavages is located at $\alpha = (270 + 180/2)°$.
- Pattern formed by V_α and V_β of a system having fourth order harmonics shows that the pattern has three cleavages. Angular separation between two consecutive cleavage (θ_C) = 120°. One of the cleavages is located at $\alpha = (270 + 120)°$.
- Pattern formed by V_α and V_β of a system having fifth order harmonics shows that the pattern has four cleavages. Angular separation between two consecutive cleavage (θ_C) = 90°. One of the cleavages is located at $\alpha = (270 + 90/2)°$.
- Pattern formed by V_α and V_β of a system having sixth order harmonics shows that the pattern has five cleavages. Angular separation between two consecutive cleavage (θ_C) = 72°. One of the cleavages is located at $\alpha = (270 + 72)°$.
- Pattern formed by V_α and V_β of a system having seventh order harmonics shows that the pattern has six cleavages. Angular separation between two consecutive cleavage (θ_C) = 60°. One of the cleavages is located at $\alpha = (270 + 60/2)°$.
- Pattern formed by V_α and V_β of a system having eighth order harmonics shows that the pattern has seven cleavages. Angular separation between two consecutive cleavage (θ_C) = 51.4°. One of the cleavages is located at $\alpha = (270 + 51.4)°$.

- Pattern formed by V_α and V_β of a system having ninth order harmonics shows that the pattern has eight cleavages. Angular separation between two consecutive cleavage $(\theta_C) = 45°$. One of the cleavages is located at $\alpha = (270 + 45/2)°$.
- Pattern formed by V_α and V_β of a system having tenth order harmonics shows that the pattern has nine cleavages. Angular separation between two consecutive cleavage $(\theta_C) = 40°$. One of the cleavages is located at $\alpha = (270 + 40)°$.
- Pattern formed by V_α and V_β of a system having eleventh order harmonics shows that the pattern has ten cleavages. Angular separation between two consecutive cleavage $(\theta_C) = 36°$. One of the cleavages is located at $\alpha = (270 + 36/2)°$.
- Pattern formed by V_α and V_β of a system having twelfth order harmonics shows that the pattern has eleven cleavages. Angular separation between two consecutive cleavage $(\theta_C) = 32.7°$. One of the cleavages is located at $\alpha = (270 + 32.7)°$.
- Pattern formed by V_α and V_β of a system having thirteenth order harmonics shows that the pattern has twelve cleavages. Angular separation between two consecutive cleavage $(\theta_C) = 30°$. One of the cleavages is located at $\alpha = (270 + 30/2)°$.

All the observations are summarized and put in Table 13.1

Patterns and Table 13.1 show that cleavages are symmetrically oriented with a particular angular separation given by α and it is observed that at least one cleavage will appear at an angle θ_C. The number of cleavages, angle α and angle θ_C follow a common relationship as given in Table 13.1. An inference can be drawn as:

1. $n = C + 1$
2. $\theta_C = 360°/C$
3. $\alpha = 270° + \theta_C/2$ for odd harmonic
4. $\alpha = 270° + \theta_C$ for even harmonic

The relationship between percentage of amplitude of harmonic and depth of locus along angle (α) is presented in Table 13.2. It shows that the depth of the cleavage is directly proportional to the percentage amplitude of the harmonic present in a system. Thus the percentage of nth order harmonic can be determined by the depth of the locus along the angle α_n.

Table 13.1 Data for C, θ_C and α

C	θ_C	α
1	–	270°
2	180°	270° + 180°/2
3	120°	270° + 120°
4	90°	270° + 90°/2
5	72°	270° + 72°
6	60°	270° + 60°/2
7	51.4°	270° + 51.4°
8	45°	270° + 45°/2
9	40°	270° + 40°
10	36°	270° + 36°/2
11	32.7°	270° + 32.7°
12	30°	270° + 30°/2

Table 13.2 CMS Rule set for harmonic assessment in Clarke plane by FPEM

Rule no	Rule
1	If cleavage appears, then there is harmonic in the system
2	If rule 1 is true and if the number of cleavages is C and order of harmonic is n, then $n = C + 1$
3	If rule 1 is true, then there will be at least one cleavage at an angle θ_C given by $\alpha_n = 270° + \theta_C/2$ for even order (n) or, $\alpha_n = 270° + \theta_C$ for odd order (n) where, $\theta_C = 360°/C$
4	If rules 1, 2 and 3 are true, then percentage amplitude of harmonic (P) is proportional to depth of locus along angle (α_n)
5	Phase angle difference of the harmonic component from the fundamental is equal to the shift of the cleavage's angle multiplied by the number of cleavages found in Clarke plane, i.e. $\varphi_n = C \times \varphi_c$ where $C = N - 1$

b. Signals Having Harmonics with no Phase Shift from the Fundamental: Using Clarke transformation matrix some patterns have been generated for system of different order of phase shifted harmonics [5]. The presence of the harmonics brings some cleavages in the elliptical pattern developed from the voltage waveforms in Clarke domain. Shift in phase angle of the harmonics with respect to the fundamental results in small shift of the angular orientation of these cleavages. Figure 13.3 shows the patterns generated by phase-shifted third order harmonics in Clarke plane.

Here the angle of each cleavage shifts by some angle (Say, ϕ_C) with the increase of phase angle difference (ϕ_n) of the harmonic component from the fundamental. From the changes of the angle of cleavages with the change of phase angle observed in Clarke plane, **CMS Rule set for harmonic assessment in Clarke plane by FPEM**

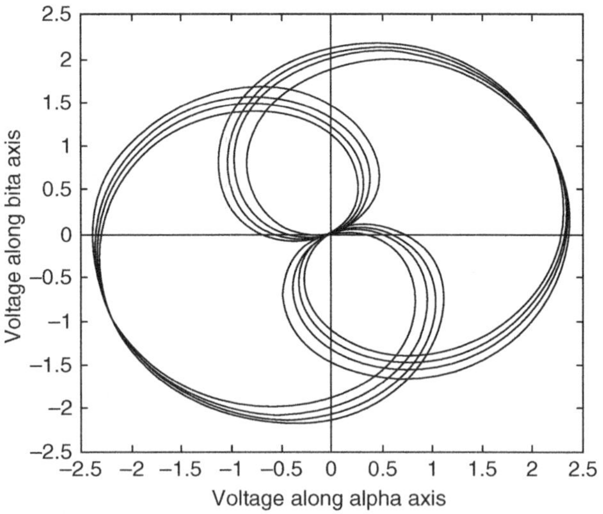

Fig. 13.3 Patterns generated by phase-shifted third order harmonics in Clarke plane

has been developed for identifying the presence of a particular harmonic present in a power system.

This rule set can be utilised for detection of presence of harmonics in system voltages and currents.

13.3 Harmonic Analysis in Park Plane

This section presents an approach for monitoring the harmonics of a power system in Park plane. Monitoring is done by developing some rules in Park domain (d-q plane) to analyze power quality in respect of harmonic contents present in the system using sampled power system data.

Using Park transformation matrix, voltages in d-q plane can be given as

$$\begin{bmatrix} v_d \\ v_q \end{bmatrix} = [Park\ Matrix] \times \begin{bmatrix} v_R \\ v_Y \\ v_B \end{bmatrix} \tag{13.12}$$

where phase voltages are given by (13.1–13.3).

In similar way d and q axis currents are given by

$$\begin{bmatrix} i_d \\ i \end{bmatrix} = [Park\ Matrix] \times \begin{bmatrix} i_R \\ i_Y \\ i_B \end{bmatrix} \tag{13.13}$$

In d-q plane, for a complete cycle, the instantaneous values of V_d and V_q collected from a balanced and harmonic free system form a circle. The presence of a harmonic brings cleavages in this circular pattern [6]. Depth of these cleavages varies with the variation of magnitude of harmonic content expressed as percent of fundamental.

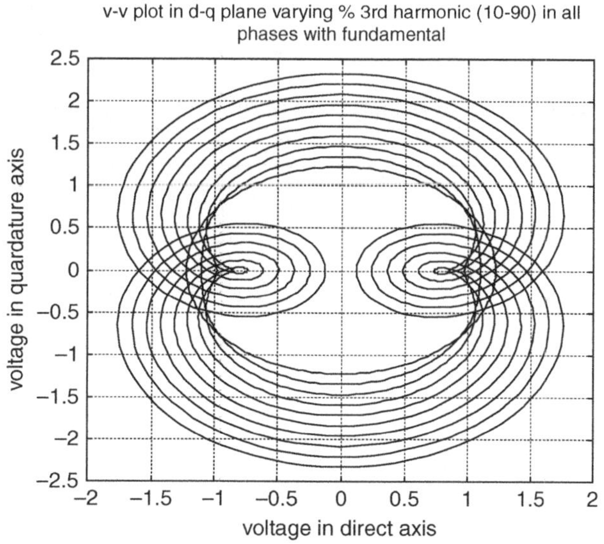

Fig. 13.4 Patterns formed by V_d and V_q in Park plane by voltage signals having increasing amplitude of third harmonic

Pattern formed by V_d and V_q of a system free from harmonics and having constant fundamental frequency for one complete cycle does not have any cleavage on it.

a. Signals Having Harmonics with no Phase Shift from the Fundamental: Patterns formed by V_d and V_q of a system consisting of third harmonic of varying amplitude as percent of fundamental is shown in Fig. 13.4. In this figure the outermost pattern corresponds to the highest percentage of third harmonic content whereas the innermost one corresponds to the lowest.

It is observed that cleavages are symmetrically oriented with a particular angular separation given by α. It is also observed that at least one cleavage will appear at an angle θ_C. α, θ_C and number of cleavages (C) follow a common relationship. This is presented in Table 13.3 where n is the order of harmonic. Table 13.3 shows that α is unique for each harmonic.

The depth (D) of a cleavage changes with the increase of P, magnitude of harmonic as percent of the fundamental component and this is presented in Table 13.4. It shows that the depth (D) of a cleavage increases with the increase of magnitude (P) of harmonic as percent of the fundamental component.

Table 13.3 Relationship among α, θ_C, order of harmonics (N) and number of cleavages (C)

N	C	θ_C	α	Remarks
2	1	–	270°	N = C + 1
3	2	180°	270° + 180°/2	$\theta_C = 360°/C$
4	3	120°	270° + 120°	$\alpha = 270° + \theta_C/2$
5	4	90°	270° + 90°/2	for odd harmonic
6	5	72°	270° + 72°	$\alpha = 270° + \theta_C$
7	6	60°	270° + 60°/2	for even harmonic
8	7	51.4°	270° + 51.4°	
9	8	45°	270° + 45°/2	
10	9	40°	270° + 40°	
11	10	36°	270° + 36°/2	
12	11	32.7°	270° + 32.7°	
13	12	30°	270° + 30°/2	

Table 13.4 Relationship between depth (D) of cleavage and amplitude of harmonic (P)

P × 100	D	Remarks
0	0	P = k × D/1.5
0.1	0.15	where k is a constant
0.2	0.30	
0.3	0.45	
0.4	0.60	
0.5	0.75	
0.6	0.90	
0.7	1.05	
0.8	1.20	
0.9	1.35	
1.0	1.5	

b. Signals Having Harmonics with Phase Shift from the Fundamental: Using Park transformation matrix some patterns have been generated for system of different order of phase shifted harmonics [5]. The presence of the harmonic brings some cleavages in the elliptical pattern developed from the voltage waveforms in Park domain. Shift in phase angle results in small shift of the angular orientation of these cleavages. Fig. 13.5 shows the patterns generated by phase-shifted third order harmonics in Park plane.

Here the angle of each cleavage shifts by some angle (say, ϕ_C) with the increase of phase angle difference (ϕ_n) of the harmonic component from the fundamental.

A rule set has been developed based on the observations made in Park plane using simulated data as presented in Table 13.5 [5, 6]. This rule set can be utilised for detection presence of harmonics in system voltages and currents.

Algorithm The algorithm for harmonic assessment using feature patterns in Park plane [7] is as follows:

1. Take power system data through instrument transformer.
2. Perform sampling of data
3. Digitize data
4. Take data through data acquisition system (DAS)
5. Normalize voltage and current waveforms
6. Transform voltage and current waveforms into Park plane.
7. Plot data in voltage-voltage and current-current planes in Park domain.
8. Generate feature patterns of different power system data.
9. Apply rule sets and make observation for harmonic assessment.

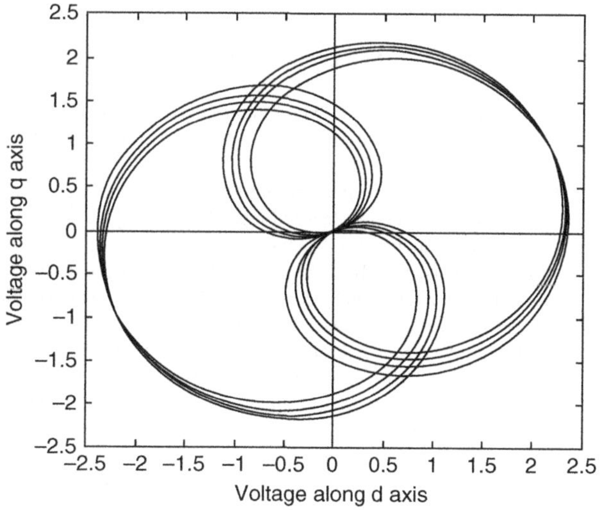

Fig. 13.5 Patterns generated by phase-shifted third order harmonics in Park plane

Table 13.5 CMS Rule set for harmonic assessment in Park plane by FPEM

Rule no	Rule
1	If cleavage appears, then there is harmonic in the system
2	If rule 1 is true and if the number of cleavages is C and order of harmonic is n, then n = C + 1
3	If rule 1 is true, then there will be at least one cleavage at an angle θ_C given by $\alpha_n = 270° + \theta_C/2$ for even order (n) or, $\alpha_n = 270° + \theta_C$ for odd order (n) where, $\theta_C = 360°/C$
4	If rules 1, 2 and 3 are true, then percentage of harmonic (P) is proportional to depth of locus along angle (α_n)
5	Phase angle difference of the harmonic component from the fundamental is equal to the shift of the cleavage's angle multiplied by the number of cleavages found in Clarke or Park plane, i.e. $\varphi_n = C \times \varphi_c$ where $C = N - 1$

13.4 Discussion

The result of this chapter shows that, the problems faced in previous chapter have been overcome in this chapter by generating patterns in Clarke and Park planes. Harmonic assessment, i.e. order and amplitude of harmonics can be determined in Clarke and Park domain. It is important to note that patterns generated in Clarke and Park domain have similar nature. Thus the rule sets obtained in Clarke and Park domain are of similar type. The algorithm developed is capable of monitoring harmonics with order, phase angle and amplitude in percent of fundamental.

References

[1] Chattopadhyay, S., Mitra, M., Sengupta, S.: Power quality assessment in V-V, Clarke and Park domain. UPEC, pp. 669–676 (2007)

[2] Chattopadhyay, S., Mitra, M., Sengupta, S.: A novel approach for harmonic analysis. Proceedings of CEE-2007-National Seminar on Creativity in Engineering in the Era of Globalization, Hooghly, 10–11th February 2007

[3] Chattopadhyay, S., Mitra, M., Sengupta, S.: Quality assessment in an electric energy system by harmonic analysis. Proceedings of ICRAMME 05- International Conference on Recent Advances in Mechanical & Material Engineering, in Kuala Lumpur, 30–31 May 2005

[4] Chattopadhyay, S., Mitra, M., Sengupta, S.: Pattern assessment of single harmonic in Clarke plane. Proceedings of MS2006-international conference on modeling and simulation by AMSE (France), in Kuala Lumpur, April 2006

[5] Chattopadhyay, S., Chakraborty, A., Das, R., Mitra, M., Sengupta, S.: Analysis of phase shifted harmonics in Clarke and Park domain power quality analysis in park plane. Proceedings of MS2006-international conference on modeling and simulation by AMSE (France), in Kuala Lumpur, April 2006

[6] Chattopadhyay, S., Mitra, M., Sengupta, S.: Power quality analysis in park plane. Best of 2006-volume on Modeling and Simulation, AMSE (France), pp. 8–16 (2006)

[7] Chattopadhyay, S., Mitra, M., Sengupta, S.: Computerized assessment of electric power quality. Proceedings of 12th State Science Congress organized by DST and Govt. of West Bengal in Kolkata, in February–March 2005

Chapter 14
Harmonic Assessment by Area Based Technique in V–V and I–I Planes

Abstract The chapter is started with an introduction of area based technique. It then describes the theory of the area based technique for harmonic assessments and relevant power quality indices. Fundamental frequency is measured them amplitude and phase angle of harmonic components are measured. Based on those parameters, total harmonic distortion factor is measured. An algorithm has been developed for application of area based technique in harmonics assessment. The chapter ends with a discussion on the technique.

14.1 Introduction

Main limitation of feature pattern extraction method is that it is unable to measure fundamental frequency, amplitudes and phases of harmonic components and harmonic distortions in voltage–voltage and current–current planes. This limitation has been overcome by using area based technique. Some reference voltage and current signals having sinusoidal nature and different frequencies have been defined. Different areas have been formed by the power system voltage and current signals and the reference signals. Areas have been calculated. Mathematical relations have been established between those areas with amplitude and phase angles of voltage and current waveforms wherefrom contributions of fundamental waveform and harmonic components have been assessed separately. Total harmonic distortion factors have also been measured.

14.2 Area Based Technique (ABT)

14.2.1 Area and Powers

In presence of harmonics, voltage and current waveforms of a power system are represented as

DOI 10.1007/978-94-007-0635-4_14,

$$v(t) = \sum_{n=1,2,3,.....n_V} V_n \sin(n\omega t - \varphi_n) \tag{14.1}$$

$$i(t) = \sum_{n=1,2,3,.....n_I} I_n \sin(n\omega t - \theta_n) \tag{14.2}$$

where, n is order of harmonics, n_V is the highest order of harmonics present in voltage waveform, n_I is the highest order of harmonics present in current waveform, ω corresponds to fundamental frequency. ω is assumed to be constant in one complete cycle. The area covered by these two waveforms in v–i plane in one complete cycle will be

$$A_{TOTAL}^{v-i} = \int vdi \quad \text{or,} \quad A_{TOTAL}^{v-i} = k_1 \sum_n n V_n I_n \sin\phi_n \tag{14.3}$$

where,
$\phi_n = \varphi_n - \theta_n$ and $k_1 = 0.1\omega$

From (14.1), sum of product of reactive power (Q_n) of each component with their order is given by

$$\sum_{n=1,2,3,...} \frac{1}{2} n V_n I_n \sin(\phi_n) = \sum_n nQ_n$$

$$\text{or,} \quad \sum_{n=1,2,3,...} \frac{1}{2} n V_n I_n \sin(\phi_n) = \frac{1}{2} K_1 A_{TOTAL}^{v-i} \tag{14.4}$$

where, $K_1 = 1/k_1$

In voltage–current vs time (vi–t) plane, the area covered by curve in presence of harmonics in one complete cycle will be

$$\begin{aligned} A_{TOTAL}^{vi-t} &= \int vidt \\ &= \int \sum_{n=1,2,3,......} V_n \sin(n\omega t - \varphi_n) \sum_{n=1,2,3,......} I_n \sin(n\omega t - \theta_n)dt \qquad (14.5) \\ &= k_2 \sum_n V_n I_n \cos\varphi_n \end{aligned}$$

where,
$\phi_n = \varphi_n - \theta_n$ and k_2 is a constant.

From (14.3) active power can be obtained as

$$\sum_{n=1,2,3,\ldots} \frac{1}{2} V_n I_n \cos(\phi_n) = \sum_n P_n$$

$$= \frac{1}{2} K_2 A_{TOTAL}^{vi-t} \tag{14.6}$$

where, $K_2 = 1/k_2$

Fundamental and harmonic components of same order, which are present in both voltage and current waveform, contribute to the active and reactive power (15.4), (15.6). (15.3) shows that area in (v–i) plane contains the information of sine of angle between the voltage and current waveform whereas, (14.5) shows that the area under the curve formed by vi in (vi–t) plane gives the information of cosine of the phase angle difference between voltage and current waveform. In the following scheme, a reference signal of sinusoidal nature has been defined and voltage (v) and current (i) signals have been plotted with the reference signal in (v–reference signal), (v. reference signal–t), (i–reference signal) and (i. reference signal–t) planes respectively, areas have been calculated wherefrom amplitude and phase angles of voltage and current signals of fundamental as well as harmonic components have been isolated [1–3].

14.2.2 Fundamental Frequency and Reference Signal for Assessment of Fundamental Component

For assessment of distortion factors of voltage and current, contribution of fundamental and harmonic components should be separated. To assess contribution of fundamental waveform of voltage, a reference signal of unity amplitude is defined with the following properties:

$$v_{REF1}(t) = \sin \omega_1 t \tag{14.7}$$

Here ω_1 corresponds to the fundamental frequency of the reference signal $v_{REF\,1}(t)$. This fundamental frequency of the reference signal ω_1 should be equal to the fundamental frequency ω of the real signal collected from the power system, which is initially an unknown quantity. To select the value of ω_1, let us first take, $\omega_1 = 50 \pm \Delta\omega$, where, $\Delta\omega$ is a small incremental value by which ω_1 differs from ω. The reference signal is then plotted against any one of the voltage signals of phase R, Y and B which have been taken from the power system data, if the angular frequency of reference signal is not equal to the angular frequency of the voltage signal of the system, then the starting and ending point will not meet each other. There will be a finite gap between these two points as shown in Fig. 14.1.

If ω_1 is changed gradually by changing $\Delta\omega$ in such a way that the starting and ending points come closer to each other then, it can be said that ω_1 is approaching

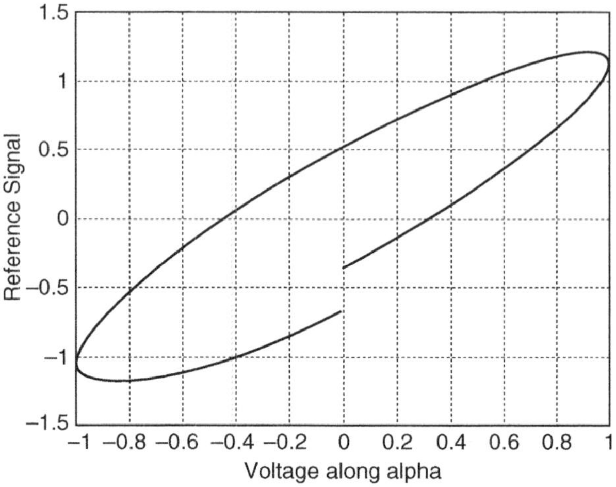

Fig. 14.1 Curves formed by $v_{REF1}(t)$ and $v_R(t)$ (in normalized and per unit form) having different fundamental frequency

towards the angular frequency (ω) which corresponds to the fundamental frequency of voltage signal of the system. When starting and ending points come at minimum distance from each other, ideally zero, then it can be said that $\omega_1 = \omega$. Then the reference signal will become

$$v_{REF1}(t) = \sin \omega t \tag{14.8}$$

It should be noted that the reference signal does not consist of harmonic components which may be in the real data. As a result of this, area formed by reference and real signals will not depend on any other frequency except fundamental.

Real voltage and current signals are plotted with the reference signal (14.8) and the areas covered by the curve in one complete cycle have been calculated wherefrom amplitude and phase angle of fundamental component, have been derived for assessment of contribution of fundamental component in distortion.

$$i_{REF1}(t) = \sin \omega t \tag{14.9}$$

14.2.3 Reference Signal for Assessment of Harmonic Components

Once the reference signals for assessment of fundamental frequency are defined, reference signals required for assessment of mth order voltage signal and n^{th} order current signal can be defined as

$$v_{REFm}(t) = \sin (m\omega t) \tag{14.10}$$

$$i_{REFn}(t) = \sin (n\omega t) \tag{14.11}$$

14.2.4 Contribution of Fundamental Component

Area covered by curve $v_R\ (t)$ and $v_{REF\,1}(t)$ waveforms in $(v_R - v_{REF})$ plane will be

$$A_{R1}^{(v_R - v_{REF})} = k_1 V_{R1} \sin \varphi_{R1}$$

$$V_{R1} \sin \varphi_{R1} = K_1 A_{R1}^{(v_R - v_{REF})} \tag{14.12}$$

Area formed by curve in $(v_{REF} v_R - \mathrm{t})$ plane will be

$$A_{R1}^{(v_{REF} v_R - t)} = k_2 V_{R1} \cos \varphi_{R1}$$

$$V_{R1} \cos \varphi_{R1} = K_2 A_{R1}^{(v_{REF}\, v_R - t)} \tag{14.13}$$

Using (14.12) and (14.13)

$$V_{R1} = \sqrt{\left\{K_1 A_{R1}^{(v_R - v_{REF})}\right\}^2 + \left\{K_2 A_{R1}^{(v_{REF}\, v_R - t)}\right\}^2} \tag{14.14}$$

$$\varphi_{R1} = \tan^{-1}\left(\frac{K_1 A_{R1}^{(v_R - v_{REF})}}{K_2 A_{R1}^{(v_{REF} v_R - t)}}\right) \tag{14.15}$$

Area covered by curve $i_R(t)$ and $i_{REF1}\ (t)$ waveforms in $(i_R$ - $i_{REF})$ plane will be

$$A_{R1}^{(i_R - i_{REF})} = k_1 I_{R1} \sin \theta_{R1}$$

$$I_{R1} \sin \theta_{R1} = K_1 A_{R1}^{(i_R - i_{REF})} \tag{14.16}$$

Area formed by curve in $(i_{REF}\, i_R - t)$ plane will be

$$A_{R1}^{(i_{REF}\, i_R - t)} = k_2 I_{R1} \cos \theta_{R1}$$

$$I_{R1} \cos \theta_{R1} = K_2 A_{R1}^{(i_{REF}\, i_R - t)} \tag{14.17}$$

Using (14.16) and (14.17)

$$I_{R1} = \sqrt{\left\{K_1 A_{R1}^{(i_R - i_{REF})}\right\}^2 + \left\{K_2 A_{R1}^{(i_{REF}\, i_R - t)}\right\}^2} \tag{14.18}$$

$$\theta_{R1} = \tan^{-1}\left(\frac{K_1 A_{R1}^{(i_R - i_{REF})}}{K_2 A_{R1}^{(i_{REF} i_R - t)}}\right) \tag{14.19}$$

Thus (14.14), (14.15) and (14.18), (14.19) give the fundamental components of voltage and current signals in magnitude and phase respectively [3]. This area based technique (ABT) has been named as **CMS Area Based Technique (CMSABT) for assessment of fundamental component**.

14.2.5 Contribution of Harmonic Components

Area covered by curve $v_R(t)$ and $v_{REFn}(t)$ waveforms in $(v_R - v_{REF})$ plane will be

$$A_{Rm}^{(v_R - v_{REF})} = k_1 V_{Rm} \sin \varphi_{Rm}$$

$$V_{Rm} \sin \varphi_{Rm} = K_1 A_{Rm}^{(v_R - v_{REF})} \tag{14.20}$$

Area formed by curve in $(v_{REF}\ v_R - t)$ plane will be

$$A_{Rm}^{(v_{REF}\ v_R - t)} = k_2 V_{Rm} \cos \varphi_{Rm}$$

$$V_{Rm} \cos \varphi_{Rm} = K_2 A_{Rm}^{(v_{REF}\ v_R - t)} \tag{14.21}$$

Using (14.20) and (14.21)

$$V_{Rm} = \sqrt{\left\{K_1 A_{Rm}^{(v_R - v_{REF})}\right\}^2 + \left\{K_2 A_{Rm}^{(v_{REF}\ v_R - t)}\right\}^2} \tag{14.22}$$

$$\varphi_{Rm} = \tan^{-1}\left(\frac{K_1 A_{Rm}^{(v_R - v_{REF})}}{K_2 A_{Rm}^{(v_{REF}\ v_R - t)}}\right) \tag{14.23}$$

Area covered by curve $i_R(t)$ and $i_{REFn}(t)$ waveforms in $(i_R - i_{REF})$ plane will be

$$A_{Rn}^{(i_R - i_{REF})} = k_1 I_{Rn} \sin \theta_{R1}$$

$$I_{Rn} \sin \theta_{Rn} = K_1 A_{Rn}^{(i_R - i_{REF})} \tag{14.24}$$

Area formed by curve in $(i_{REF}\ i_R - t)$ plane will be

$$A_{Rn}^{(i_{REF}\ i_R - t)} = k_2 I_{Rn} \cos \theta_{R1}$$

$$I_{Rn} \cos \theta_{Rn} = K_2 A_{Rn}^{(i_{REF}\ i_R - t)} \tag{14.25}$$

Using (14.24) and (14.25)

$$I_{Rn} = \sqrt{\left\{K_1 A_{Rn}^{(i_R - i_{REF})}\right\}^2 + \left\{K_2 A_{Rn}^{(i_{REF}\ i_R - t)}\right\}^2} \tag{14.26}$$

$$\theta_{Rn} = \tan^{-1}\left(\frac{K_1 A_{Rn}^{(i_R - i_{REF})}}{K_2 A_{Rn}^{(i_{REF}\ i_R - t)}}\right) \tag{14.27}$$

Thus (14.22) and (14.23) give the magnitude and phase of mth harmonic voltage signal whereas (14.26) and (14.27) give nth harmonic current in phase and magnitude. This area based technique (ABT) is called as **CMS Area Based Technique (CMSABT) for assessment of harmonic component. CMS equations for total harmonic distortion factors are discussed in the following sections.**

14.2.6 CMS Equations for Total Harmonic Distortion Factors

Total harmonic distortion factor [1, 4] for voltage in individual phase (say, R phase) is written as using (14.14) and (14.22)

$$THD_{RV} = \frac{\sqrt{\sum\limits_{m=2,3,\ldots\ldots\ldots).} (V_{Rm})^2}}{V_{R1}}$$

$$= \frac{\sqrt{\sum\limits_{m=2,3,\ldots\ldots\ldots} \left[\left\{K_1 A_{Rm}^{(v_R - v_{REF})}\right\}^2 + \left\{K_2 A_{Rm}^{(v_{REF}\, v_R - t)}\right\}^2\right]}}{\sqrt{\left[\left\{K_1 A_{R1}^{(v_R - v_{REF})}\right\}^2 + \left\{K_2 A_{R1}^{(v_{REF} v_R - t)}\right\}^2\right]}} \tag{14.28}$$

Total harmonic distortion factor for current in individual phase (say, R phase) is written as using (14.18) and (14.26)

$$THD_{RI} = \frac{\sqrt{\sum\limits_{n=2,3,\ldots\ldots\ldots} (I_{Rn})^2}}{I_{R1}}$$

$$= \frac{\sqrt{\sum\limits_{n=2,3,\ldots\ldots\ldots} \left[\left\{K_1 A_{Rn}^{(i_R - i_{REF})}\right\}^2 + \left\{K_2 A_{Rn}^{(i_{REF}\, i_R - t)}\right\}^2\right]}}{\sqrt{\left[\left\{K_1 A_{R1}^{(i_R - i_{REF})}\right\}^2 + \left\{K_2 A_{R1}^{(i_{REF}\, i_R - t)}\right\}^2\right]}} \tag{14.29}$$

In the name of the introducer of the equations, (14.28 and 14.29) are known as CMS equations of total harmonic distortions of voltage and current signals.

14.3 Algorithm

The analyzer works on an algorithm developed based on the principles discussed in Sect. 14.2. The algorithm has been explained with the help of a flow diagram shown in Fig. 14.2. First fundamental frequency is determined. Then reference signal is set and different loops are to be formed. Areas are calculated to assess fundamental as well as harmonic components. After this, total harmonic distortions are calculated.

14.4 Discussion

In this chapter area based technique has been applied for harmonic assessment in single phase signals. By this technique, fundamental frequency, its amplitude along with amplitude and phase angle of harmonic components can be individually

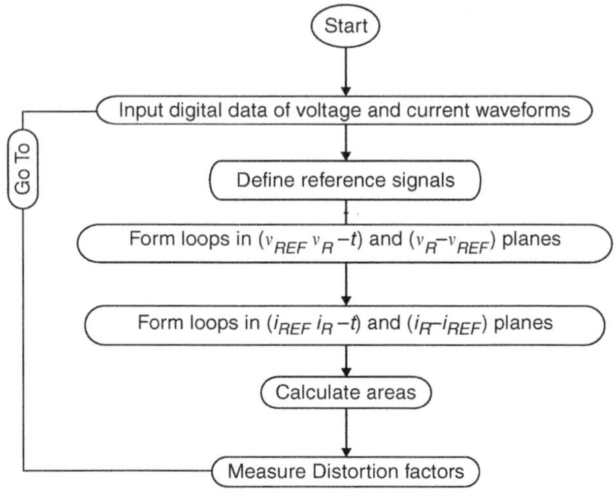

Fig. 14.2 Flow diagram for analyzer

determined. This makes the technique very useful for determination of total harmonic distortion. Error in measurement by this technique is very small.

References

[1] Chattopadhyay, S., Mitra, M., Sengupta, S.: Area based mathematical model for harmonic distortion. POWERCON 2008, International Conference on power system organized by IEEE in New Delhi, 12–15 October 2008

[2] Chattopadhyay, S., Mitra, M., Sengupta, S.: Area based approach in power quality assessment. Int. J. Power Manage. Electron. (2008)

[3] Chattopadhyay, S., Mitra, M., Sengupta, S.: Single cycle mathematical model for harmonic power. Proceedings of MS'07-India: International Conference on Modeling and Simulation in Kolkata, 3–5 December 2007

[4] Chattopadhyay, S., Mitra, M., Sengupta, S.: An area based approach for total harmonic distortion analysis. Proceedings of International Conference of Emerging Trends in Electrical Engineering Kolkata, 12–14 January 2007

Chapter 15
Harmonic Assessment by Area Based Technique in Clarke and Park Planes

Abstract This chapter starts with an introduction mentioning the limitation of feature pattern extraction method in harmonics assessment in Clarke and Park planes. It then develops the mathematics for harmonic assessment in Clarke and Park planes by area based technique. Fundamental frequency is measured and then amplitude and phase angle of harmonic components are measured in Clarke and Park planes. Based on those parameters, total harmonic distortion factors are measured. An algorithm has been developed for application of area based technique in harmonics assessment. The chapter ends with a discussion.

15.1 Introduction

Feature pattern extraction method applied in Clarke and Park plane is not capable of assessing amplitude and phase angle of individual harmonic components and their effects along different axes in Clarke and Park planes. Moreover, it was difficult to measure total harmonic distortion factors. These limitations have been overcome in this chapter by applying an area based approach for assessment of harmonic distortion in Clarke and Parke planes. Three phase power system data of voltage and current signals have been transformed into Clarke and Park planes. Some specific reference signals having sinusoidal nature and different frequencies have been defined. Areas formed by the real power system data with the reference signal have been calculated [1]. Mathematical relations have been established between those areas with amplitude and phase angles of voltage and current waveforms wherefrom individual contributions of fundamental waveform and harmonic components have been assessed separately [2–4]. Total harmonic distortion factor of the three phase power system has been measured.

S. Chattopadhyay et al., *Electric Power Quality*, Power Systems,
DOI 10.1007/978-94-007-0635-4_15,

15.2 Voltage and Current in Clarke ($\alpha-\beta$) Plane

Let, voltages in matrix form are

$$\left[v_{R,Y,B}\right] = \begin{bmatrix} \sum\limits_{m=1,2,3,\ldots\ldots m_V} V_{Rm} \sin(m\omega t - \varphi_{Rm}) \\ \sum\limits_{m=1,2,3,\ldots\ldots m_V} V_{Ym} \sin(m\omega t - 2\pi/3 - \varphi_{Ym}) \\ \sum\limits_{m=1,2,3,\ldots\ldots m_V} V_{Bm} \sin(m\omega t - 4\pi/3 - \varphi_{Bm}) \end{bmatrix} \tag{15.1}$$

Line currents in matrix form can be written as

$$\left[i_{R,Y,B}\right] = \begin{bmatrix} \sum\limits_{n=1,2,3,\ldots\ldots n_I} I_{Rn} \sin(n\omega t - \theta_{Rn}) \\ \sum\limits_{n=1,2,3,\ldots\ldots n_I} I_{Yn} \sin(n\omega t - 2\pi/3 - \theta_{Yn}) \\ \sum\limits_{n=1,2,3,\ldots\ldots n_I} I_{Bn} \sin(n\omega t - 4\pi/3 - \theta_{Bn}) \end{bmatrix} \tag{15.2}$$

Voltages and currents in Clarke Plane ($\alpha-\beta$ plane) are given by

$$\left[v_{\alpha,\beta,0}\right] = \begin{bmatrix} v_\alpha \\ v_\beta \\ v_0 \end{bmatrix} = \begin{bmatrix} \sum\limits_{m=1,2,3\ldots.m_V} V_{\alpha m} \sin(m\omega t - \varphi_{\alpha m}) \\ \sum\limits_{m=1,2,3\ldots.m_V} V_{\beta m} \sin(m\omega t - \varphi_{\beta m}) \\ v_0 \end{bmatrix} \tag{15.3}$$

$$\left[i_{\alpha,\beta,0}\right] = \begin{bmatrix} i_\alpha \\ i_\beta \\ i_0 \end{bmatrix} = \begin{bmatrix} \sum\limits_{n=1,2,3\ldots.n_I} I_{\alpha n} \sin(n\omega t - \theta_{\alpha n}) \\ \sum\limits_{n=1,2,3\ldots.n_I} I_{\beta n} \sin(n\omega t - \theta_{\beta n}) \\ i_0 \end{bmatrix} \tag{15.4}$$

Well known Clarke transformation matrix is given as

$$[CM] = Clarke\ Matrix \tag{15.5}$$

Voltage and current in Clarke plane can be obtained from multiplying phase voltages and line current by Clarke Matrix; i.e.,

$$\left[v_{\alpha,\beta,0}\right] = [CM] \times \left[v_{R,Y,B}\right] \tag{15.6}$$

$$\left[i_{\alpha,\beta,0}\right] = [CM] \times \left[i_{R,Y,B}\right] \tag{15.7}$$

15.3 Reference Signal for Assessment of Fundamental Component

As discussed in previous chapter, reference signals may be defined as

$$v_{REF1}(t) = \sin \omega t \tag{15.8}$$

$$i_{REF1}(t) = \sin \omega t \tag{15.9}$$

$$v_{REFm}(t) = \sin (m\omega t) \tag{15.10}$$

$$i_{REFn}(t) = \sin (n\omega t) \tag{15.11}$$

(15.8) and (15.9) are reference signals required for assessment of fundamental components of voltage and current respectively. Similarly, (15.10) and (15.11) are reference signals required for assessment of mth order harmonic component of voltage and n^{th} order harmonic component of current respectively.

15.4 Fundamental Components in Clarke Plane

Area covered by curve $v_\alpha(t)$ and $v_{REF1}(t)$ waveforms in $(v_\alpha - v_{REF})$ plane will be

$$\begin{aligned} A_{\alpha 1}^{(v_\alpha - v_{REF})} &= k_1 V_{\alpha 1} \sin \varphi_{\alpha 1} \\ V_{\alpha 1} \sin \varphi_{\alpha 1} &= K_1 A_{\alpha 1}^{(v_\alpha - v_{REF})} \end{aligned} \tag{15.12}$$

Area formed by curve in $(v_{REF} v_\alpha - t)$ plane will be

$$\begin{aligned} A_{\alpha 1}^{(v_{REF} v_\alpha - t)} &= k_2 V_{\alpha 1} \cos \varphi_{\alpha 1} \\ V_{\alpha 1} \cos \varphi_{\alpha 1} &= K_2 A_{\alpha 1}^{(v_{REF} v_\alpha - t)} \end{aligned} \tag{15.13}$$

Using (15.12) and (15.13)

$$V_{\alpha 1} = \sqrt{\left\{K_1 A_{\alpha 1}^{(v_R - v_{REF})}\right\}^2 + \left\{K_2 A_{\alpha 1}^{(v_{REF} v_\alpha - t)}\right\}^2} \tag{15.14}$$

$$\varphi_{\alpha 1} = \tan^{-1}\left(\frac{K_1 A_{\alpha 1}^{(v_\alpha - v_{REF})}}{K_2 A_{\alpha 1}^{(v_{REF} v_\alpha - t)}}\right) \tag{15.15}$$

Equations (15.14) and (15.15) represent amplitude and phase angle of fundamental component of voltage along α axis.

Area covered by curve $i_\alpha(t)$ and $i_{REF1}(t)$ waveforms in $(i_\alpha - i_{REF})$ plane will be

$$\begin{aligned} A_{\alpha 1}^{(i_\alpha - i_{REF})} &= k_1 I_{\alpha 1} \sin \theta_{\alpha 1} \\ I_{\alpha 1} \sin \theta_{\alpha 1} &= K_1 A_{\alpha 1}^{(i_\alpha - i_{REF})} \end{aligned} \tag{15.16}$$

Area formed by curve in $(i_{REF}i_{\alpha} - t)$ plane will be

$$A_{\alpha 1}^{(i_{REF}\, i_{\alpha}-t)} = k_2 I_{\alpha 1} \cos\theta_{\alpha 1}$$
$$I_{\alpha 1} \cos\theta_{\alpha 1} = K_2 A_{\alpha 1}^{(i_{REF}\, i_{\alpha}-t)} \tag{15.17}$$

Using (15.16) and (15.17)

$$I_{\alpha 1} = \sqrt{\left\{K_1 A_{\alpha 1}^{(i_{\alpha}-i_{REF})}\right\}^2 + \left\{K_2 A_{\alpha 1}^{(i_{REF}\, i_{\alpha}-t)}\right\}^2} \tag{15.18}$$

$$\theta_{\alpha 1} = \tan^{-1}\left(\frac{K_1 A_{\alpha 1}^{(i_{\alpha}-i_{REF})}}{K_2 A_{\alpha 1}^{(i_{REF}\, i_{\alpha}-t)}}\right) \tag{15.19}$$

Equations (15.18) and (15.19) represent amplitude and phase angle of fundamental component of current along α axis.

Area covered by curve $v_{\beta}(t)$ and $v_{REF1}(t)$ waveforms in $(v_{\beta} - v_{REF})$ plane will be

$$A_{\beta 1}^{(v_{\beta}-v_{REF})} = k_1 V_{\beta 1} \sin\varphi_{\beta 1}$$
$$V_{\beta 1} \sin\varphi_{\beta 1} = K_1 A_{\beta 1}^{(v_{\beta}-v_{REF})} \tag{15.20}$$

Area formed by curve in $(v_{REF}v_{\beta} - t)$ plane will be

$$A_{\beta 1}^{(v_{REF}v_{\beta}-t)} = k_2 V_{\beta 1} \cos\varphi_{\beta 1}$$
$$V_{\beta 1} \cos\varphi_{\beta 1} = K_2 A_{\beta 1}^{(v_{REF}\, v_{\beta}-t)} \tag{15.21}$$

Using (15.20) and (15.21)

$$V_{\beta 1} = \sqrt{\left\{K_1 A_{\beta 1}^{(v_{\beta}-v_{REF})}\right\}^2 + \left\{K_2 A_{\beta 1}^{(v_{REF}\, v_{\beta}-t)}\right\}^2} \tag{15.22}$$

$$\varphi_{\beta 1} = \tan^{-1}\left(\frac{K_1 A_{\beta 1}^{(v_{\beta}-v_{REF})}}{K_2 A_{\beta 1}^{(v_{REF}\, v_{\beta}-t)}}\right) \tag{15.23}$$

Equations (15.22) and (15.23) represent amplitude and phase angle of fundamental component of voltage along β axis.

Area covered by curve $i_{\beta}(t)$ and $i_{REF1}(t)$ waveforms in $(i_{\beta} - i_{REF})$ plane will be

$$A_{\beta 1}^{(i_{\beta}-i_{REF})} = k_1 I_{\beta 1} \sin\theta_{\beta 1}$$
$$I_{\beta 1} \sin\theta_{\beta 1} = K_1 A_{\beta 1}^{(i_{\beta}-i_{REF})} \tag{15.24}$$

Area formed by curve in $(i_{REF}i_{\beta} - t)$ plane will be

$$A_{\beta 1}^{(i_{REF}i_{\beta}-t)} = k_2 I_{\beta 1} \cos\theta_{\beta 1}$$
$$I_{\beta 1} \cos\theta_{\beta 1} = K_2 A_{\beta 1}^{(i_{REF}\, i_{\beta}-t)} \tag{15.25}$$

Using (15.24) and (15.25)

$$I_{\beta 1} = \sqrt{\left\{K_1 A_{\beta 1}^{(i_\beta - i_{REF})}\right\}^2 + \left\{K_2 A_{\beta 1}^{(i_{REF} i_\beta - t)}\right\}^2} \tag{15.26}$$

$$\theta_{\beta 1} = \tan^{-1}\left(\frac{K_1 A_{\beta 1}^{(i_\beta - i_{REF})}}{K_2 A_{\beta 1}^{(i_{REF}\, i_\beta - t)}}\right) \tag{15.27}$$

Equations (15.26) and (15.27) represent amplitude and phase angle of fundamental component of current along β axis.

Thus (15.14), (15.15), (15.18) and (15.19) express fundamental components of voltage and current along α-axis in magnitude and phase whereas (15.22), (15.23), (15.26) and (15.27) express fundamental components of voltage and current along β-axis of the Clarke plane. This approach is called **CMS Area Based Technique (CMSABT) for assessment of fundamental component in Clarke Plane.**

15.5 Harmonic Components in Clarke Plane

Area covered by curve $v_\alpha(t)$ and $v_{REFn}(t)$ waveforms in $(v_\alpha - v_{REF})$ plane will be

$$A_{\alpha n}^{(v_\alpha - v_{REF})} = k_1 V_{\alpha n} \sin \varphi_{\alpha n}$$

$$V_{\alpha m} \sin \varphi_{\alpha m} = K_1 A_{\alpha m}^{(v_\alpha - v_{REF})} \tag{15.28}$$

Area formed by curve in $(v_{REF} v_\alpha - t)$ plane will be

$$A_{\alpha m}^{(v_{REF}\, v_\alpha - t)} = k_2 V_{\alpha m} \cos \varphi_{\alpha m}$$

$$V_{\alpha m} \cos \varphi_{\alpha m} = K_2 A_{\alpha m}^{(v_{REF}\, v_\alpha - t)} \tag{15.29}$$

Using (15.28) and (15.29)

$$V_{\alpha m} = \sqrt{\left\{K_1 A_{\alpha m}^{(v_\alpha - v_{REF})}\right\}^2 + \left\{K_2 A_{\alpha m}^{(v_{REF}\, v_\alpha - t)}\right\}^2} \tag{15.30}$$

$$\varphi_{\alpha m} = \tan^{-1}\left(\frac{K_1 A_{\alpha m}^{(v_\alpha - v_{REF})}}{K_2 A_{\alpha m}^{(v_{REF}\, v_\alpha - t)}}\right) \tag{15.31}$$

Equations (15.30) and (15.31) represent amplitude and phase angle of harmonic component of voltage along α axis.

Area covered by curve $i_\alpha(t)$ and $i_{REFn}(t)$ waveforms in $(i_\alpha - i_{REF})$ plane will be

$$A_{\alpha n}^{(i_\alpha - i_{REF})} = k_1 I_{\alpha n} \sin \theta_{\alpha 1}$$

$$I_{\alpha n} \sin \theta_{\alpha n} = K_1 A_{\alpha n}^{(i_\alpha - i_{REF})} \tag{15.32}$$

Area formed by curve in $(i_{REF}i_{\alpha} - t)$ plane will be

$$
\begin{aligned}
A_{\alpha n}^{(i_{REF}\,i_{\alpha}-t)} &= k_2 I_{\alpha n} \cos\theta_{\alpha n} \\
I_{\alpha n} \cos\theta_{\alpha n} &= K_2 A_{\alpha n}^{(i_{REF}\,i_{\alpha}-t)}
\end{aligned} \tag{15.33}
$$

Using (15.32) and (15.33)

$$
I_{\alpha n} = \sqrt{\left\{K_1 A_{\alpha n}^{(i_{\alpha}-i_{REF})}\right\}^2 + \left\{K_2 A_{\alpha n}^{(i_{REF}\,i_{\alpha}-t)}\right\}^2} \tag{15.34}
$$

$$
\theta_{\alpha n} = \tan^{-1}\left(\frac{K_1 A_{\alpha n}^{(i_{\alpha}-i_{REF})}}{K_2 A_{\alpha n}^{(i_{REF}\,i_{\alpha}-t)}}\right) \tag{15.35}
$$

Equations (15.34) and (15.35) represent amplitude and phase angle of harmonic component of current along α axis.

Area covered by curve $v_{\beta}(t)$ and $v_{REFn}(t)$ waveforms in $(v_{\beta} - v_{REF})$ plane will be

$$
\begin{aligned}
A_{\beta n}^{(v_{\beta}-v_{REF})} &= k_1 V_{\beta n} \sin\varphi_{\beta n} \\
V_{\beta m} \sin\varphi_{\beta m} &= K_1 A_{\beta m}^{(v_{\beta}-v_{REF})}
\end{aligned} \tag{15.36}
$$

Area formed by curve in $(v_{REF}v_{\beta} - t)$ plane will be

$$
\begin{aligned}
A_{\beta m}^{(v_{REF}\,v_{\beta}-t)} &= k_2 V_{\beta m} \cos\varphi_{\beta m} \\
V_{\beta m} \cos\varphi_{\beta m} &= K_2 A_{\beta m}^{(v_{REF}\,v_{\beta}-t)}
\end{aligned} \tag{15.37}
$$

Using (15.36) and (15.37)

$$
V_{\beta m} = \sqrt{\left\{K_1 A_{\beta m}^{(v_{\beta}-v_{REF})}\right\}^2 + \left\{K_2 A_{\beta m}^{(v_{REF}\,v_{\beta}-t)}\right\}^2} \tag{15.38}
$$

$$
\varphi_{\beta m} = \tan^{-1}\left(\frac{K_1 A_{\beta m}^{(v_{\beta}-v_{REF})}}{K_2 A_{\beta m}^{(v_{REF}\,v_{\beta}-t)}}\right) \tag{15.39}
$$

Equations (15.38) and (15.39) represent amplitude and phase angle of harmonic component of voltage along β axis.

Area covered by curve $i_{\beta}(t)$ and $i_{REFn}(t)$ waveforms in $(i_{\beta} - i_{REF})$ plane will be

$$
\begin{aligned}
A_{\beta n}^{(i_{\beta}-i_{REF})} &= k_1 I_{\beta n} \sin\theta_{\beta n} \\
I_{\beta n} \sin\theta_{\beta n} &= K_1 A_{\beta n}^{(i_{\beta}-i_{REF})}
\end{aligned} \tag{15.40}
$$

Area formed by curve in $(i_{REF}i_{\beta} - t)$ plane will be

$$
A_{\beta n}^{(i_{REF}\,i_{\beta}-t)} = k_2 I_{\beta n} \cos\theta_{\beta 1}
$$

$$I_{\beta n}\cos\theta_{\beta n} = K_2 A_{\beta n}^{(i_{REF}\, i_\beta - t)} \tag{15.41}$$

Using (15.40) and (15.41)

$$I_{\beta n} = \sqrt{\left\{K_1 A_{\beta n}^{(i_\beta - i_{REF})}\right\}^2 + \left\{K_2 A_{\beta n}^{(i_{REF}\, i_\beta - t)}\right\}^2} \tag{15.42}$$

$$\theta_{\beta n} = \tan^{-1}\left(\frac{K_1 A_{\beta n}^{(i_\alpha - i_{REF})}}{K_2 A_{\beta n}^{(i_{REF}\, i_\beta - t)}}\right) \tag{15.43}$$

Equations (15.42) and (15.43) represent amplitude and phase angle of harmonic component of current along β axis.

Thus harmonic components of voltage and currents along α- and β-axes in magnitude and phase in the Clarke plane are obtained as (15.30), (15.31), (15.34), (15.35), (15.38), (15.39), (15.42) and (15.43). This technique is called as **CMS Area Based Technique (CMSABT) for assessment of harmonic component in Clarke Plane.**

15.6 CMS Equations for Total Harmonic Distortion in Clarke Plane

Total harmonic distortion for voltage along α axis in Clarke plane is written using (15.14) and (15.30) as,

$$THD_{V_\alpha} = \frac{\sqrt{\sum\limits_{m=2,3,\ldots\ldots\ldots\ldots} (V_{\alpha m})^2}}{V_{\alpha 1}} = \frac{\sqrt{\sum\limits_{m=2,3,\ldots\ldots\ldots} \left[\left\{K_1 A_{\alpha m}^{(v_\alpha - v_{REF})}\right\}^2 + \left\{K_2 A_{\alpha m}^{(v_{REF}\, v_\alpha - t)}\right\}^2\right]}}{\sqrt{\left[\left\{K_1 A_{\alpha 1}^{(v_\alpha - v_{REF})}\right\}^2 + \left\{K_2 A_{\alpha 1}^{(v_{REF}\, v_\alpha - t)}\right\}^2\right]}} \tag{15.44}$$

Total harmonic distortion factor for current along α axis in Clarke plane is written using (15.18) and (15.34) as,

$$THD_{I_\alpha} = \frac{\sqrt{\sum\limits_{n=2,3,\ldots\ldots\ldots} (I_{\alpha n})^2}}{I_{\alpha 1}} = \frac{\sqrt{\sum\limits_{n=2,3,\ldots\ldots\ldots} \left[\left\{K_1 A_{\alpha n}^{(i_\alpha - i_{REF})}\right\}^2 + \left\{K_2 A_{\alpha n}^{(i_{REF}\, i_\alpha - t)}\right\}^2\right]}}{\sqrt{\left[\left\{K_1 A_{\alpha 1}^{(i_\alpha - i_{REF})}\right\}^2 + \left\{K_2 A_{\alpha 1}^{(i_{REF}\, i_\alpha - t)}\right\}^2\right]}} \tag{15.45}$$

Total harmonic distortion factor for voltage along β axis in Clarke plane is written using (15.22) and (15.38) as,

$$THD_{V_\beta} = \frac{\sqrt{\sum\limits_{m=2,3,\ldots\ldots\ldots} (V_{\beta m})^2}}{V_{\beta 1}} = \frac{\sqrt{\sum\limits_{m=2,3,\ldots\ldots\ldots} \left[\left\{K_1 A_{\beta m}^{(v_\beta - v_{REF})}\right\}^2 + \left\{K_2 A_{\beta m}^{(v_{REF}\, v_\beta - t)}\right\}^2\right]}}{\sqrt{\left[\left\{K_1 A_{\beta 1}^{(v_\beta - v_{REF})}\right\}^2 + \left\{K_2 A_{\beta 1}^{(v_{REF}\, v_\beta - t)}\right\}^2\right]}} \quad (15.46)$$

Total harmonic distortion factor for current along β axis in Clarke plane is written using (15.26) and (15.42) as,

$$THD_{I_\beta} = \frac{\sqrt{\sum\limits_{n=2,3,\ldots\ldots\ldots} (I_{\beta n})^2}}{I_{\beta 1}} = \frac{\sqrt{\sum\limits_{n=2,3,\ldots\ldots\ldots} \left[\left\{K_1 A_{\beta n}^{(i_\beta - i_{REF})}\right\}^2 + \left\{K_2 A_{\beta n}^{(i_{REF}\, i_\beta - t)}\right\}^2\right]}}{\sqrt{\left[\left\{K_1 A_{\beta 1}^{(i_\beta - i_{REF})}\right\}^2 + \left\{K_2 A_{\beta 1}^{(i_{REF}\, i_\beta - t)}\right\}^2\right]}} \quad (15.47)$$

Equations (15.44–15.47) are known as CMS equations of total harmonic distortions in Clarke Plane [5].

15.7 Voltages and Currents in Park (d–q) Plane

The approach in Park plane [6] is almost similar to that used for analysis in Clarke plane. In presence of harmonics, voltages and currents in Park Domain can be derived from phase voltages and currents multiplying by Park transformation matrix as follows:

$$\left[v_{d,q,0}\right] = \begin{bmatrix} v_d \\ v_q \\ v_0 \end{bmatrix} = \begin{bmatrix} \sum\limits_{m=1,2,3\ldots.m_V} V_{dm} \sin\left(m\omega t - \varphi_{dm}\right) \\ \sum\limits_{m=1,2,3\ldots.m_V} V_{qm} \sin\left(m\omega t - \varphi_{qm}\right) \\ v_0 \end{bmatrix} \quad (15.48)$$

$$\left[i_{d,q,0}\right] = \begin{bmatrix} i_d \\ i_q \\ i_0 \end{bmatrix} = \begin{bmatrix} \sum\limits_{n=1,2,3\ldots.n_I} I_{dn} \sin\left(n\omega t - \theta_{dn}\right) \\ \sum\limits_{n=1,2,3\ldots.n_I} I_{qn} \sin\left(n\omega t - \theta_{qn}\right) \\ i_0 \end{bmatrix} \quad (15.49)$$

Well known Park transformation matrix is given as

$$[PM] = \sqrt{\frac{2}{3}} \begin{bmatrix} \cos\theta & \cos\left(\theta - \frac{2\pi}{3}\right) & \cos\left(\theta - \frac{4\pi}{3}\right) \\ -\sin\theta & -\sin\left(\theta - \frac{2\pi}{3}\right) & -\sin\left(\theta - \frac{4\pi}{3}\right) \\ \frac{1}{2} & \frac{1}{2} & \frac{1}{2} \end{bmatrix} \tag{15.50}$$

Voltage and current in Park plane can be obtained as

$$\left[v_{d,q,0}\right] = [PM] \times \left[v_{R,Y,B}\right] \tag{15.51}$$

$$\left[i_{d,q,0}\right] = [PM] \times \left[i_{R,Y,B}\right] \tag{15.52}$$

15.8 Reference Signal in Park Plane

Like analysis in Clarke plane, reference signals are defined in the following manner:

a. The reference signal for assessment of fundamental component of voltage waveform

$$v_{REF1}(t) = \sin\omega t \tag{15.53}$$

b. The reference signal for assessment of fundamental component of current waveform

$$i_{REF1}(t) = \sin\omega t \tag{15.54}$$

c. The reference signal for assessment of mth order harmonic component of voltage waveform

$$v_{REFm}(t) = \sin(m\omega t) \tag{15.55}$$

d. The reference signal for assessment of nth order harmonic component of current waveform

$$i_{REFn}(t) = \sin(n\omega t) \tag{15.56}$$

15.9 Fundamental Components in Park Plane

Area covered by curve $v_d(t)$ and $v_{REF1}(t)$ waveforms in $(v_d - v_{REF})$ plane will be

$$
\begin{aligned}
A_{d1}^{(v_d - v_{REF})} &= k_1 V_{d1} \sin \varphi_{d1} \\
V_{d1} \sin \varphi_{d1} &= K_1 A_{d1}^{(v_d - v_{REF})}
\end{aligned}
\tag{15.57}
$$

Area formed by curve in $(v_{REF} v_d - t)$ plane will be

$$
\begin{aligned}
A_{d1}^{(v_{REF}\, v_d - t)} &= k_2 V_{d1} \cos \varphi_{d1} \\
V_{d1} \cos \varphi_{d1} &= K_2 A_{d1}^{(v_{REF}\, v_d - t)}
\end{aligned}
\tag{15.58}
$$

Using (15.57) and (15.58)

$$
V_{d1} = \sqrt{\left\{K_1 A_{d1}^{(v_d - v_{REF})}\right\}^2 + \left\{K_2 A_{d1}^{(v_{REF}\, v_d - t)}\right\}^2}
\tag{15.59}
$$

$$
\varphi_{d1} = \tan^{-1}\left(\frac{K_1 A_{d1}^{(v_d - v_{REF})}}{K_2 A_{d1}^{(v_{REF}\, v_d - t)}}\right)
\tag{15.60}
$$

Equations (15.59) and (15.60) represent amplitude and phase angle of fundamental component of voltage along d axis.

Area covered by curve $i_d(t)$ and $i_{REF1}(t)$ waveforms in $(i_d - i_{REF})$ plane will be

$$
\begin{aligned}
A_{d1}^{(i_d - i_{REF})} &= k_1 I_{d1} \sin \theta_{d1} \\
I_{d1} \sin \theta_{d1} &= K_1 A_{d1}^{(i_d - i_{REF})}
\end{aligned}
\tag{15.61}
$$

Area formed by curve in $(i_{REF} i_d - t)$ plane will be

$$
\begin{aligned}
A_{d1}^{(i_{REF}\, i_d - t)} &= k_2 I_{d1} \cos \theta_{d1} \\
I_{d1} \cos \theta_{d1} &= K_2 A_{d1}^{(i_{REF}\, i_d - t)}
\end{aligned}
\tag{15.62}
$$

Using (15.61) and (15.62)

$$
I_{d1} = \sqrt{\left\{K_1 A_{d1}^{(i_d - i_{REF})}\right\}^2 + \left\{K_2 A_{d1}^{(i_{REF}\, i_d - t)}\right\}^2}
\tag{15.63}
$$

$$
\theta_{d1} = \tan^{-1}\left(\frac{K_1 A_{d1}^{(i_d - i_{REF})}}{K_2 A_{d1}^{(i_{REF}\, i_d - t)}}\right)
\tag{15.64}
$$

Equations (15.63) and (15.64) represent amplitude and phase angle of fundamental component of current along d axis [6].

Area covered by curve $v_q(t)$ and $v_{REF1}(t)$ waveforms in $(v_q - v_{REF})$ plane will be

$$
A_{q1}^{(v_q - v_{REF})} = k_1 V_{q1} \sin \varphi_{q1}
$$

$$V_{q1}\sin\varphi_{q1} = K_1 A_{q1}^{(v_q - v_{REF})} \tag{15.65}$$

Area formed by curve in $(v_{REF}v_q - t)$ plane will be

$$A_{q1}^{(v_{REF}\,v_q - t)} = k_2 V_{q1}\cos\varphi_{q1}$$

$$V_{q1}\cos\varphi_{q1} = K_2 A_{q1}^{(v_{REF}\,v_q - t)} \tag{15.66}$$

Using (15.65) and (15.66)

$$V_{q1} = \sqrt{\left\{K_1 A_{q1}^{(v_q - v_{REF})}\right\}^2 + \left\{K_2 A_{q1}^{(v_{REF}\,v_q - t)}\right\}^2} \tag{15.67}$$

$$\varphi_{q1} = \tan^{-1}\left(\frac{K_1 A_{q1}^{(v_q - v_{REF})}}{K_2 A_{q1}^{(v_{REF}\,v_q - t)}}\right) \tag{15.68}$$

Equations (15.67) and (15.68) represent amplitude and phase angle of fundamental component of voltage along q axis.

Area covered by curve $i_q(t)$ and $i_{REF1}(t)$ waveforms in $(i_q - i_{REF})$ plane will be

$$A_{q1}^{(i_q - i_{REF})} = k_1 I_{q1}\sin\theta_{q1}$$

$$I_{q1}\sin\theta_{q1} = K_1 A_{q1}^{(i_q - i_{REF})} \tag{15.69}$$

Area formed by curve in $(i_{REF}i_q - t)$ plane will be

$$A_{q1}^{(i_{REF}i_q - t)} = k_2 I_{q1}\cos\theta_{q1}$$

$$I_{q1}\cos\theta_{q1} = K_2 A_{q1}^{(i_{REF}i_q - t)} \tag{15.70}$$

Using (15.69) and (15.70)

$$I_{q1} = \sqrt{\left\{K_1 A_{q1}^{(i_q - i_{REF})}\right\}^2 + \left\{K_2 A_{q1}^{(i_{REF}\,i_q - t)}\right\}^2} \tag{15.71}$$

$$\theta_{q1} = \tan^{-1}\left(\frac{K_1 A_{q1}^{(i_q - i_{REF})}}{K_2 A_{q1}^{(i_{REF}\,i_q - t)}}\right) \tag{15.72}$$

Equations (15.71) and (15.72) represent amplitude and phase angle of fundamental component of current along q axis [6].

This area based technique (ABT) is named as **CMS Area Based Technique (CMSABT) for assessment of fundamental component in Park Plane.**

15.10 Harmonic Components in Park Plane

Area covered by curve $v_d(t)$ and $v_{REFn}(t)$ waveforms in $(v_d - v_{REF})$ plane will be

$$\begin{aligned} A_{dn}^{(v_d - v_{REF})} &= k_1 V_{dn} \sin \varphi_{dn} \\ V_{dm} \sin \varphi_{dm} &= K_1 A_{dm}^{(v_d - v_{REF})} \end{aligned} \tag{15.73}$$

Area formed by curve in $(v_{REF} v_d - t)$ plane will be

$$\begin{aligned} A_{dm}^{(v_{REF}\, v_d - t)} &= k_2 V_{dm} \cos \varphi_{dm} \\ V_{dm} \cos \varphi_{dm} &= K_2 A_{dm}^{(v_{REF}\, v_d - t)} \end{aligned} \tag{15.74}$$

Using (15.73) and (15.74)

$$V_{dm} = \sqrt{\left\{K_1 A_{dm}^{(v_d - v_{REF})}\right\}^2 + \left\{K_2 A_{dm}^{(v_{REF}\, v_d - t)}\right\}^2} \tag{15.75}$$

$$\varphi_{dm} = \tan^{-1}\left(\frac{K_1 A_{dm}^{(v_d - v_{REF})}}{K_2 A_{dm}^{(v_{REF}\, v_d - t)}}\right) \tag{15.76}$$

Equations (15.75) and (15.76) represent amplitude and phase angle of harmonic component of voltage along d axis.

Area covered by curve $i_d(t)$ and $i_{REFn}(t)$ waveforms in $(i_d - i_{REF})$ plane will be

$$\begin{aligned} A_{dn}^{(i_d - i_{REF})} &= k_1 I_{dn} \sin \theta_{dn} \\ I_{dn} \sin \theta_{dn} &= K_1 A_{dn}^{(i_d - i_{REF})} \end{aligned} \tag{15.77}$$

Area formed by curve in $(i_{REF} i_d - t)$ plane will be

$$\begin{aligned} A_{dn}^{(i_{REF}\, i_d - t)} &= k_2 I_{dn} \cos \theta_{dn} \\ I_{dn} \cos \theta_{dn} &= K_2 A_{dn}^{(i_{REF}\, i_d - t)} \end{aligned} \tag{15.78}$$

Using (15.77) and (15.78)

$$I_{dn} = \sqrt{\left\{K_1 A_{dn}^{(i_d - i_{REF})}\right\}^2 + \left\{K_2 A_{dn}^{(i_{REF}\, i_d - t)}\right\}^2} \tag{15.79}$$

$$\theta_{dn} = \tan^{-1}\left(\frac{K_1 A_{dn}^{(i_d - i_{REF})}}{K_2 A_{dn}^{(i_{REF}\, i_d - t)}}\right) \tag{15.80}$$

Equations (15.79) and (15.80) represent amplitude and phase angle of harmonic component of current along d axis [6].

Area covered by curve $v_q(t)$ and $v_{REFn}(t)$ waveforms in $(v_q - v_{REF})$ plane will be

$$A_{qn}^{(v_q - v_{REF})} = k_1 V_{qn} \sin \varphi_{qn}$$

$$V_{qm} \sin \varphi_{qm} = K_1 A_{qm}^{(v_q - v_{REF})} \tag{15.81}$$

Area formed by curve in $(v_{REF} v_q - t)$ plane will be

$$A_{qm}^{(v_{REF}\, v_q - t)} = k_2 V_{qm} \cos \varphi_{qm}$$

$$V_{qm} \cos \varphi_{qm} = K_2 A_{qm}^{(v_{REF}\, v_q - t)} \tag{15.82}$$

Using (15.81) and (15.82)

$$V_{qm} = \sqrt{\left\{K_1 A_{qm}^{(v_q - v_{REF})}\right\}^2 + \left\{K_2 A_{qm}^{(v_{REF}\, v_q - t)}\right\}^2} \tag{15.83}$$

$$\varphi_{qm} = \tan^{-1}\left(\frac{K_1 A_{qm}^{(v_q - v_{REF})}}{K_2 A_{qm}^{(v_{REF}\, v_q - t)}}\right) \tag{15.84}$$

Equations (15.83) and (15.84) represent amplitude and phase angle of harmonic component of voltage along q axis.

Area covered by curve $i_q(t)$ and $i_{REFn}(t)$ waveforms in $(i_q - i_{REF})$ plane will be

$$A_{qn}^{(i_q - i_{REF})} = k_1 I_{qn} \sin \theta_{qn}$$

$$I_{qn} \sin \theta_{qn} = K_1 A_{qn}^{(i_q - i_{REF})} \tag{15.85}$$

Area formed by curve in $(i_{REF} i_q - t)$ plane will be

$$A_{qn}^{(i_{REF} i_q - t)} = k_2 I_{qn} \cos \theta_{qn}$$

$$I_{qn} \cos \theta_{qn} = K_2 A_{qn}^{(i_{REF}\, i_q - t)} \tag{15.86}$$

Using (15.85) and (15.86)

$$I_{qn} = \sqrt{\left\{K_1 A_{qn}^{(i_q - i_{REF})}\right\}^2 + \left\{K_2 A_{qn}^{(i_{REF}\, i_q - t)}\right\}^2} \tag{15.87}$$

$$\theta_{qn} = \tan^{-1}\left(\frac{K_1 A_{qn}^{(i_q - i_{REF})}}{K_2 A_{qn}^{(i_{REF}\, i_q - t)}}\right) \tag{15.88}$$

Equations (15.87) and (15.88) represent amplitude and phase angle of harmonic component of current along q axis [6].

The above technique is called as **CMS Area Based Technique (CMSABT) for assessment of harmonic component in Park Plane.**

15.11 CMS Equations for Total Harmonic Distortion Factors

Total harmonic distortion factor for voltage along d axis

$$THD_{V_d} = \frac{\sqrt{\sum\limits_{m=2,3,\ldots\ldots\ldots} (V_{dm})^2}}{V_{d1}} = \frac{\sqrt{\sum\limits_{m=2,3,\ldots\ldots\ldots} \left[\left\{K_1 A_{dm}^{(v_d - v_{REF})}\right\}^2 + \left\{K_2 A_{dm}^{(v_{REF}\, v_d - t)}\right\}^2\right]}}{\sqrt{\left[\left\{K_1 A_{d1}^{(v_d - v_{REF})}\right\}^2 + \left\{K_2 A_{d1}^{(v_{REF}\, v_d - t)}\right\}^2\right]}} \tag{15.89}$$

Total harmonic distortion factor for current along d-axis is written as

$$THD_{I_d} = \frac{\sqrt{\sum\limits_{n=2,3,\ldots\ldots\ldots} (I_{dn})^2}}{I_{d1}} = \frac{\sqrt{\sum\limits_{n=2,3,\ldots\ldots\ldots} \left[\left\{K_1 A_{dn}^{(i_d - i_{REF})}\right\}^2 + \left\{K_2 A_{dn}^{(i_{REF}\, i_d - t)}\right\}^2\right]}}{\sqrt{\left[\left\{K_1 A_{d1}^{(i_d - i_{REF})}\right\}^2 + \left\{K_2 A_{d1}^{(i_{REF}\, i_d - t)}\right\}^2\right]}} \tag{15.90}$$

Total harmonic distortion factor for voltage along q axis

$$THD_{V_q} = \frac{\sqrt{\sum\limits_{m=2,3,\ldots\ldots\ldots} (V_{qm})^2}}{V_{q1}} = \frac{\sqrt{\sum\limits_{m=2,3,\ldots\ldots\ldots} \left[\left\{K_1 A_{qm}^{(v_q - v_{REF})}\right\}^2 + \left\{K_2 A_{qm}^{(v_{REF}\, v_q - t)}\right\}^2\right]}}{\sqrt{\left[\left\{K_1 A_{q1}^{(v_q - v_{REF})}\right\}^2 + \left\{K_2 A_{q1}^{(v_{REF}\, v_q - t)}\right\}^2\right]}} \tag{15.91}$$

Total harmonic distortion factor for current along q axis

$$THD_{I_q} = \frac{\sqrt{\sum\limits_{n=2,3,\ldots\ldots\ldots} (I_{qn})^2}}{I_{q1}} = \frac{\sqrt{\sum\limits_{n=2,3,\ldots\ldots\ldots} \left[\left\{K_1 A_{qn}^{(i_q - i_{REF})}\right\}^2 + \left\{K_2 A_{qn}^{(i_{REF}\, i_q - t)}\right\}^2\right]}}{\sqrt{\left[\left\{K_1 A_{q1}^{(i_q - i_{REF})}\right\}^2 + \left\{K_2 A_{q1}^{(i_{REF}\, i_q - t)}\right\}^2\right]}} \tag{15.92}$$

Equations (15.89–15.92) are known as CMS equations of total harmonic distortions in Park Plane.

15.12 Discussion

In this chapter area based technique has been applied for harmonic assessment in Clarke and Park planes. In three-phase power system consisting of three voltage waveforms and three current waveforms, use of the Clarke–Parke domain consisting of two voltage waveforms and two current waveforms, has reduced the computational task to a great extent. Thus the above area based approach and Clarke–Park Domain for three phase system may be good platform for power quality parameter estimation.

References

[1] Chattopadhyay, S., Mitra, M., Sengupta, S.: Area based approach in power quality assessment. Int. J. Power Manage. Electron. (2008)
[2] Chattopadhyay, S., Mitra, M., Sengupta, S.: Area based mathematical model for harmonic distortion. POWERCON 2008, International Conference on power system organized by IEEE, New Delhi, 12–15 October 2008
[3] Chattopadhyay, S., Mitra, M., Sengupta, S.: Area based approach for harmonic analysis. Proceedings of 14th Science Congress organized by DST and Government of West Bengal, Kolkata, February–March 2007
[4] Chattopadhyay, S., Mitra, M., Sengupta, S.: Single cycle mathematical model for harmonic power. Proceedings of MS'07-India: International Conference on Modeling and Simulation, Kolkata, 3–5 December 2007
[5] Chattopadhyay, S., Mitra, M., Sengupta, S.: Area based approach for three phase power quality assessment in Clarke plane. J. Electrical Syst. 4(3), (2008)
[6] Chattopadhyay, S., Mitra, M., Sengupta, S.: Harmonic analysis in a three-phase system using park transformation technique. Modeling-A, AMSE Int. J. Model. Simul. 80(3), France (2007)

Chapter 16
Assessment of Power Components by FPEM and ABT

Abstract In this chapter different power components are measured using feature pattern extraction method and area based technique. Conditions for minimum active and reactive power, power factor and direction of power flow are developed. Active and reactive power contributed by fundament components are measured by area based technique. The contribution of harmonic components in active and reactive power is assessed. Based on these, different power distortion factors are calculated. For measurement of three phase power components, Clarke and Park Planes are considered.

16.1 Introduction

Power components are assessed here by feature pattern extraction method (FPEM) and area based techniques (ABT) [1]. Inability of FPEM is overcome by area based technique. Using FPEM, maxima or minima of active power, reactive power, power factor angle and direction of power flow in a power system are monitored. Then different powers contributed by fundamental as well as harmonic components are measured by area based techniques, wherefrom different power distortion factors are calculated [2–4].

16.2 Power Components by FPEM

A good quality power consists of balanced voltage and current waveforms free from harmonics. Voltages can be represented as

$$v_R(t) = V_R \sin \omega t \tag{16.1}$$

$$v_Y(t) = V_Y \sin (\omega t + 120^\circ) \tag{16.2}$$

$$v_B(t) = V_B \sin (\omega t + 240^\circ) \tag{16.3}$$

S. Chattopadhyay et al., *Electric Power Quality,* Power Systems,
DOI 10.1007/978-94-007-0635-4_16, © Springer Science+Business Media B.V. 2011

Similarly, currents can be represented as

$$i_R(t) = I_R \sin(\omega t + \phi) \tag{16.4}$$

$$i_Y(t) = I_Y \sin(\omega t + 120^\circ + \phi) \tag{16.5}$$

$$i_B(t) = I_B \sin(\omega t + 240^\circ + \phi) \tag{16.6}$$

For a balanced system,

$$V_R = V_Y = V_B \quad \text{and} \quad I_R = I_Y = I_B$$

From (16.1) to (16.6), it can be written as

$$v_Y(t) = V_Y \sin\left(\sin^{-1}\frac{v_R(t)}{V_R} + 120^\circ\right) \tag{16.7}$$

$$v_B(t) = V_B \sin\left(\sin^{-1}\frac{v_Y(t)}{V_Y} + 120^\circ\right) \tag{16.8}$$

$$v_R(t) = V_R \sin\left(\sin^{-1}\frac{v_B(t)}{V_B} - 240^\circ\right) \tag{16.9}$$

$$i_Y(t) = I_Y \sin\left(\sin^{-1}\frac{i_R(t)}{I_R} + 120^\circ\right) \tag{16.10}$$

$$i_B(t) = I_B \sin\left(\sin^{-1}\frac{i_Y(t)}{I_Y} + 120^\circ\right) \tag{16.11}$$

$$i_R(t) = I_R \sin\left(\sin^{-1}\frac{i_B(t)}{I_B} - 240^\circ\right) \tag{16.12}$$

(16.7), (16.8) and (16.9) represent three ellipses in voltage–voltage plane. (16.10), (16.11) and (16.12) represent three ellipses in current–current plane. In an unbalanced system V_R, V_Y, V_B and I_R, I_Y, I_B will not be equal, i.e.,

$$V_R \neq V_Y \neq V_B \quad \text{and} \quad I_R \neq I_Y \neq I_B$$

In normalized form, voltages can be written as

$$v_R^N(t) = \frac{v_R(t)}{V_R} = \sin\omega t \tag{16.13}$$

$$v_Y^N(t) = \frac{v_Y(t)}{V_R} = \frac{V_Y}{V_R}\sin(\omega t + 120^\circ) \tag{16.14}$$

$$v_B^N(t) = \frac{v_B(t)}{V_R} = \frac{V_B}{V_R}\sin(\omega t + 240^\circ) \tag{16.15}$$

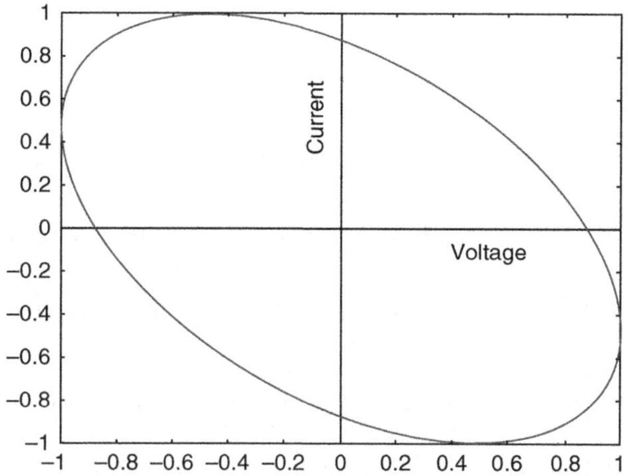

Fig. 16.1 Elliptical pattern in voltage–current plane

Similarly, currents can be represented as

$$i_R^N(t) = \frac{i_R(t)}{I_R} = \sin(\omega t + \phi) \tag{16.16}$$

$$i_Y^N(t) = \frac{i_Y(t)}{I_R} = \frac{I_Y}{I_R}\sin(\omega t + 120^\circ + \phi) \tag{16.17}$$

$$i_B^N(t) = \frac{i_B(t)}{I_R} = \frac{I_B}{I_R}\sin(\omega t + 240^\circ + \phi) \tag{16.18}$$

One such pattern formed by voltage and current signals has been shown in Fig. 16.1. The unbalance in a system changes the length of major and minor axes.

Let,

$$v(t) = V\sin\omega t \tag{16.19}$$

and

$$i(t) = I\sin(\omega t + \phi) \tag{16.20}$$

If v and i are plotted in x–y plane, an ellipse will be formed, area of which will be given by

$$\begin{aligned} A_E &= \int v\,di \\ &= \int V\sin\omega t\;di \end{aligned}$$

Now,

$$di = I\omega\cos(\omega t + \phi)\,dt$$

Therefore,

$$\begin{aligned}
A_E &= \int V \sin \omega t \; I\omega \cos(\omega t + \phi)\, dt \\
&= VI\omega \int \sin \omega t \cos(\omega t + \phi)\, dt \\
&= VI\omega \int \sin \omega t\, (\cos \omega t \cos \phi + \sin \omega t \sin \phi)\, dt \\
&= VI\omega \left(\cos \phi \int \sin \omega t \cos \omega t \; dt + \sin \phi \int \sin \omega t \sin \omega t \; dt \right) \\
&= VI\omega \left\{ \frac{1}{2} \cos \phi \int \sin 2\omega t \; dt + \frac{1}{2} \sin \phi \int (1 - \cos 2\omega t)\, dt \right\}
\end{aligned} \tag{16.21}$$

(Taking the limit of t from 0 to 0.02 ms for 50 Hz system)

$$\begin{aligned}
A_E &= VI\omega \left(0 + \frac{1}{2} \sin \phi \times 0.02 \right) \\
&= 0.01\omega VI \sin \phi \\
&= K \times \text{Reactive power}
\end{aligned} \tag{16.22}$$

where,

$$K = 0.01\omega \tag{16.23}$$

To measure three phase reactive power, let,

V_{RY} = amplitude of R-phase to Y-phase voltage
V_{YB} = amplitude of Y-phase to B-phase voltage
I_L = line current

Therefore, areas representing the powers will be

$$A_{RY} = KV_{RY} I_L \sin(\phi + 30°) = \sqrt{3}\, KV_L I_L \sin(\phi + 30°) \tag{16.24}$$

$$A_{YB} = KV_{YB} I_L \sin(\phi - 30°) = \sqrt{3}\, KV_L I_L \sin(\phi - 30°) \tag{16.25}$$

$$A_T = A_{RY} + A_{YB} = \sqrt{3}\, KV_L I_L \sin(\phi + 30°) + \sqrt{3}\, KV_L I_L \sin(\phi - 30°) \tag{16.26}$$

$$\begin{aligned}
A_T &= \sqrt{3} KV_L I_L \times 2 \times \sin \phi \cos 30° \\
&= 3KV_L I_L \sin \phi \\
&= \text{K} \times \text{three phase reactive power}
\end{aligned} \tag{16.27}$$

To extract the features from the patterns, the following parameters are introduced as indicated in Fig. 16.1:

$$\text{at } X = X_{MIN},\ Y = Y_1 \tag{16.28}$$

$$\text{at } X = X_{MAX},\ Y = Y_2 \tag{16.29}$$

$$\text{at } Y = Y_{MIN},\ X = X_1 \tag{16.30}$$

$$\text{at } Y = Y_{MAX},\ X = X_2 \tag{16.31}$$

$$\text{at } X = 0,\ |Y| = Y_0 \tag{16.32}$$

$$\text{at } Y = 0,\ |X| = X_0 \tag{16.33}$$

$$y = Y_1 \sim Y_2 \tag{16.34}$$

$$x = X_1 \sim X_2 \tag{16.35}$$

Now, from (16.19) and (16.20)
$v(t)$ is maximum when $\sin \omega t = 1$
i.e., $\omega t = 90°$
From (16.29)

$$\begin{aligned} Y_2 &= i(t)_{v(t)=\text{maximum}} \\ &= i(t)_{\omega t=90°} \\ &= I \sin(90° + \phi) \\ &= I \cos\phi \\ &= \cos\phi \\ &[if\ I = 1] \end{aligned} \tag{16.36}$$

$Y = i(t)$ is maximum when $\sin(\omega t + \phi) = 1$
i.e., $\omega t = 90° - \phi$
From (16.31)

$$\begin{aligned} X_2 &= v(t)_{i(t)=\text{maximum}} \\ &= v(t)_{\omega t=90°-\phi} \\ &= V \sin(90° - \phi) \\ &= V \cos\phi \\ &= \cos\phi \\ &[if\ I = 1] \end{aligned} \tag{16.37}$$

$$X_2 = V \sin(\omega t) = \sin(90° - \phi) \tag{16.38}$$

Therefore

$$X_2 = Y_2 = \cos\phi \tag{16.39}$$

$v(t)$ is zero when $\sin \omega t = 0$

i.e., $\omega t = 0°$

From (16.32)

$$\begin{aligned} Y_0 &= i(t)_{v(t)=0} \\ &= i(t)_{\omega t=0°} \\ &= I \sin(0° + \phi) \\ &= I \sin \phi \\ &= \sin \phi \\ &[if\ I = 1] \end{aligned} \tag{16.40}$$

Case I—Condition of Maximum Reactive Power and Minimum Active Power So in normalized waveform, condition of maximum reactive power and minimum active power is [5]

$$\begin{gathered} Y_0 = \sin \phi = 1 \quad and \quad X_2 = Y_2 = 0 \\ [if\ I = V = 1] \end{gathered} \tag{16.41}$$

During this condition, the ellipse becomes a circle of area

$$A_E = K\ VI \sin 90° = K \tag{16.42}$$

Case II—Condition of Minimum Reactive Power and Maximum Active Power Similarly, in normalized waveform, condition of minimum reactive power and maximum active power is [5]

$$\begin{gathered} Y_0 = \sin \phi = 0 \quad and \quad X_2 = Y_2 = 1 \\ [if\, I = V = 1] \end{gathered} \tag{16.43}$$

During this condition, the ellipse becomes a straight line of area

$$A_E = K\ VI \sin 0° = 0 \tag{16.44}$$

16.3 CMS Rule Set for Power Components by FPEM [5, 6]

Rule 1

$$X_2 = Y_2 = \cos \phi, \quad \sin \phi = \surd\left(1 - X_2{}^2\right), \quad \phi = \cos^{-1} X_2 \tag{16.45}$$

Rule 2

Sign change of $(X_2 - X_1)$ means change in direction of active power flow as shown in Table 16.1.

Table 16.1 Detection of direction of power flow

$X_2 - X_1$ (v-i plane)	Verification
Positive	Power flow is from source end to load end
Negative	Power flow is from load end to source end

Rule 3

Condition of Maximum Reactive Power and Minimum Active Power: In normalized waveform, condition of maximum reactive power and minimum active power is

$$Y_0 = \sin\phi = 1 \quad and \quad X_2 = Y_2 = 0 \\ [if\, I = V = 1] \tag{16.46}$$

Rule 4

In normalized waveform, condition of minimum reactive power and maximum active power is

$$Y_0 = \sin\phi = 0 \quad and \quad X_2 = Y_2 = 1 \\ [if\ I = V = 1] \tag{16.47}$$

During this condition, the ellipse becomes a straight line of area

$$A_E = K\ VI \sin 0° = 0 \tag{16.48}$$

16.4 Limitations of CMS Rule Set for Power Components by FPEM

The rule set developed in this chapter is capable of measuring active and reactive power. It can measure power factor. It can indicate whether maximum or minimum reactive or active power has been reached. But, main limitation of this method is that the rule set is not capable of giving information of powers contributed by harmonic components. Thus assessment of contribution of harmonics on active and reactive powers is not possible in this method.

16.5 Power Component Assessment by Area Based Technique

The limitations have been over come in the following sections using area based approach. Mathematical expressions [2, 4, 6, 7] for amplitude and phase angles of voltage and current waveforms have been derived and briefly discussed in Chaps. 8 and 9. Using those expressions, contributions of fundamental waveform and harmonic components to different power component have been assessed separately. Active power, reactive power and total power contributed by fundamental as well as harmonic components have been calculated [1–4, 7]. Distortion factors in respect of these powers have also been measured for individual phase as well as in Clarke and Park planes.

16.6 Power Components of R, Y and B Phases

Power drawn by a three phase system is primarily shared by all the three phases and contribution of each phase consists of the contributions of (a) the fundamental components of voltage and current and (b) the harmonic components of voltage and current.

16.6.1 Contribution of Fundamental Components

In the following analysis contribution of only R phase has been considered. Amplitude and phase angles of fundamental component of voltage in phase R is obtained from (14.14) and (14.15) as

$$V_{R1} = \sqrt{\left\{K_1 A_{R1}^{(v_R - v_{REF})}\right\}^2 + \left\{K_2 A_{R1}^{(v_{REF}\, v_R - t)}\right\}^2} \tag{16.49}$$

$$\varphi_{R1} = \tan^{-1}\left(\frac{K_1 A_{R1}^{(v_R - v_{REF})}}{K_2 A_{R1}^{(v_{REF}\, v_R - t)}}\right) \tag{16.50}$$

Amplitude and phase angles of fundamental component of current in phase R is obtained from (14.18) and (14.19) as

$$I_{R1} = \sqrt{\left\{K_1 A_{R1}^{(i_R - i_{REF})}\right\}^2 + \left\{K_2 A_{R1}^{(i_{REF}\, i_R - t)}\right\}^2} \tag{16.51}$$

$$\theta_{R1} = \tan^{-1}\left(\frac{K_1 A_{R1}^{(i_R - i_{REF})}}{K_2 A_{R1}^{(i_{REF}\, i_R - t)}}\right) \tag{16.52}$$

Using (16.49) to (16.52), reactive and active power contributed by fundamental component can be expressed as

$$\begin{aligned} Q_{R1} &= \frac{1}{2} V_{R1} I_{R1} \sin(\varphi_{R1} - \theta_{R1}) \\ &= \frac{1}{2}\sqrt{\left\{K_1 A_{R1}^{(v_R - v_{REF})}\right\}^2 + \left\{K_2 A_{R1}^{(v_{REF}\, v_R - t)}\right\}^2} \\ &\times \sqrt{\left\{K_1 A_{R1}^{(i_R - i_{REF})}\right\}^2 + \left\{K_2 A_{R1}^{(i_{REF}\, i_R - t)}\right\}^2} \\ &\times \sin\left(\tan^{-1}\left(\frac{K_1 A_{R1}^{(v_R - v_{REF})}}{K_2 A_{R1}^{(v_{REF}\, v_R - t)}}\right) - \tan^{-1}\left(\frac{K_1 A_{R1}^{(i_R - i_{REF})}}{K_2 A_{R1}^{(i_{REF}\, i_R - t)}}\right)\right) \end{aligned} \tag{16.53}$$

$$\begin{aligned}
P_{R1} &= \frac{1}{2} V_{R1} I_{R1} \cos(\varphi_{R1} - \theta_{R1}) \\
&= \frac{1}{2}\sqrt{\left\{K_1 A_{R1}^{(v_R - v_{REF})}\right\}^2 + \left\{K_2 A_{R1}^{(v_{REF}\, v_R - t)}\right\}^2} \\
&\times \sqrt{\left\{K_1 A_{R1}^{(i_R - i_{REF})}\right\}^2 + \left\{K_2 A_{R1}^{(i_{REF}\, i_R - t)}\right\}^2} \\
&\times \cos\left(\tan^{-1}\left(\frac{K_1 A_{R1}^{(v_R - v_{REF})}}{K_2 A_{R1}^{(v_{REF}\, v_R - t)}}\right) - \tan^{-1}\left(\frac{K_1 A_{R1}^{(i_R - i_{REF})}}{K_2 A_{R1}^{(i_{REF}\, i_R - t)}}\right)\right)
\end{aligned} \tag{16.54}$$

Total power contributed by fundamental components of voltage and current in phase R can be written as

$$S_{R1} = P_{R1} + jQ_{R1} \tag{16.55}$$

Similarly, power contribution of fundamental components of Y and B phases can be obtained.

16.6.2 Contribution of Harmonic Components

Amplitude and phase angles of harmonic component of mth harmonic component of voltage in phase R is obtained from (14.22) and (14.23) as

$$V_{Rm} = \sqrt{\left\{K_1 A_{Rm}^{(v_R - v_{REF})}\right\}^2 + \left\{K_2 A_{Rm}^{(v_{REF}\, v_R - t)}\right\}^2} \tag{16.56}$$

$$\varphi_{Rm} = \tan^{-1}\left(\frac{K_1 A_{Rm}^{(v_R - v_{REF})}}{K_2 A_{Rm}^{(v_{REF}\, v_R - t)}}\right) \tag{16.57}$$

Amplitude and phase angles of harmonic component of nth harmonic component of current in phase R is obtained from (14.26) and (14.27) as

$$I_{Rn} = \sqrt{\left\{K_1 A_{Rn}^{(i_R - i_{REF})}\right\}^2 + \left\{K_2 A_{Rn}^{(i_{REF}\, i_R - t)}\right\}^2} \tag{16.58}$$

$$\theta_{Rn} = \tan^{-1}\left(\frac{K_1 A_{Rn}^{(i_R - i_{REF})}}{K_2 A_{Rn}^{(i_{REF}\, i_R - t)}}\right) \tag{16.59}$$

Only, harmonics of voltage and current of same order will have contribution to the power components. Hence considering harmonics of voltage current of same order (i.e., n = m), from (16.56) to (16.59), reactive and active power contributed by harmonic component can be expressed as

$$
\begin{aligned}
Q_{Rn} &= \frac{1}{2} V_{Rn} I_{Rn} \sin(\varphi_{Rn} - \theta_{Rn}) \\
&= \frac{1}{2} \sqrt{\left\{K_1 A_{Rn}^{(v_R - v_{REF})}\right\}^2 + \left\{K_2 A_{Rn}^{(v_{REF}\, v_R - t)}\right\}^2} \\
&\quad \times \sqrt{\left\{K_1 A_{Rn}^{(i_R - i_{REF})}\right\}^2 + \left\{K_2 A_{Rn}^{(i_{REF}\, i_R - t)}\right\}^2} \\
&\quad \times \sin\left(\tan^{-1}\left(\frac{K_1 A_{Rn}^{(v_R - v_{REF})}}{K_2 A_{Rn}^{(v_{REF}\, v_R - t)}}\right) - \tan^{-1}\left(\frac{K_1 A_{Rn}^{(i_R - i_{REF})}}{K_2 A_{Rn}^{(i_{REF}\, i_R - t)}}\right)\right)
\end{aligned}
\tag{16.60}
$$

$$
\begin{aligned}
P_{Rn} &= \frac{1}{2} V_{Rn} I_{Rn} \cos(\varphi_{Rn} - \theta_{Rn}) \\
&= \frac{1}{2} \sqrt{\left\{K_1 A_{Rn}^{(v_R - v_{REF})}\right\}^2 + \left\{K_2 A_{Rn}^{(v_{REF}\, v_R - t)}\right\}^2} \\
&\quad \times \sqrt{\left\{K_1 A_{Rn}^{(i_R - i_{REF})}\right\}^2 + \left\{K_2 A_{Rn}^{(i_{REF}\, i_R - t)}\right\}^2} \\
&\quad \times \cos\left(\tan^{-1}\left(\frac{K_1 A_{Rn}^{(v_R - v_{REF})}}{K_2 A_{Rn}^{(v_{REF}\, v_R - t)}}\right) - \tan^{-1}\left(\frac{K_1 A_{Rn}^{(i_R - i_{REF})}}{K_2 A_{Rn}^{(i_{REF}\, i_R - t)}}\right)\right)
\end{aligned}
\tag{16.61}
$$

Total power contributed by harmonic components of voltage and current in phase R can be written as

$$
S_{Rn} = \sum_{n=2,3,\ldots\ldots\ldots\ldots} (P_{Rn} + jQ_{Rn}) \tag{16.62}
$$

Total power contributed both by fundamental and harmonics of voltage and current in phase R can be written as

$$
S_R = \sum_{n=1,2,3,\ldots\ldots\ldots\ldots} (P_{Rn} + jQ_{Rn}) \tag{16.63}
$$

Similarly, contributions by Y and B phases can be calculated.

16.7 Power Components in Clarke Plane [8]

16.7.1 Contribution of Fundamental Components

Amplitude and phase angle of fundamental component of voltage along α axis in Clarke Plane is obtained from (15.14) and (15.15) as

$$
V_{\alpha 1} = \sqrt{\left\{K_1 A_{\alpha 1}^{(v_R - v_{REF})}\right\}^2 + \left\{K_2 A_{\alpha 1}^{(v_{REF}\, v_\alpha - t)}\right\}^2} \tag{16.64}
$$

$$\varphi_{\alpha 1} = \tan^{-1}\left(\frac{K_1 A_{\alpha 1}^{(v_\alpha - v_{REF})}}{K_2 A_{\alpha 1}^{(v_{REF}\, v_\alpha - t)}}\right) \tag{16.65}$$

Amplitude and phase angle of fundamental component of current along α axis in Clarke Plane is obtained from (15.18) and (15.19) as

$$I_{\alpha 1} = \sqrt{\left\{K_1 A_{\alpha 1}^{(i_\alpha - i_{REF})}\right\}^2 + \left\{K_2 A_{\alpha 1}^{(i_{REF}\, i_\alpha - t)}\right\}^2} \tag{16.66}$$

$$\theta_{\alpha 1} = \tan^{-1}\left(\frac{K_1 A_{\alpha 1}^{(i_\alpha - i_{REF})}}{K_2 A_{\alpha 1}^{(i_{REF}\, i_\alpha - t)}}\right) \tag{16.67}$$

Using (16.64) to (16.67), reactive and active power contributed by fundamental components along α axis can be expressed as

$$\begin{aligned} Q_{\alpha 1} &= \frac{1}{2} V_{\alpha 1} I_{\alpha 1} \sin(\varphi_{\alpha 1} - \theta_{\alpha 1}) \\ &= \frac{1}{2}\sqrt{\left\{K_1 A_{\alpha 1}^{(v_\alpha - v_{REF})}\right\}^2 + \left\{K_2 A_{\alpha 1}^{(v_{REF}\, v_\alpha - t)}\right\}^2} \\ &\times \sqrt{\left\{K_1 A_{\alpha 1}^{(i_\alpha - i_{REF})}\right\}^2 + \left\{K_2 A_{\alpha 1}^{(i_{REF}\, i_\alpha - t)}\right\}^2} \\ &\times \sin\left(\tan^{-1}\left(\frac{K_1 A_{\alpha 1}^{(v_\alpha - v_{REF})}}{K_2 A_{\alpha 1}^{(v_{REF}\, v_\alpha - t)}}\right) - \tan^{-1}\left(\frac{K_1 A_{\alpha 1}^{(i_\alpha - i_{REF})}}{K_2 A_{\alpha 1}^{(i_{REF}\, i_\alpha - t)}}\right)\right) \end{aligned} \tag{16.68}$$

$$\begin{aligned} P_{\alpha 1} &= \frac{1}{2} V_{\alpha 1} I_{\alpha 1} \cos(\varphi_{\alpha 1} - \theta_{\alpha 1}) \\ &= \frac{1}{2}\sqrt{\left\{K_1 A_{\alpha 1}^{(v_\alpha - v_{REF})}\right\}^2 + \left\{K_2 A_{\alpha 1}^{(v_{REF}\, v_\alpha - t)}\right\}^2} \\ &\times \sqrt{\left\{K_1 A_{\alpha 1}^{(i_\alpha - i_{REF})}\right\}^2 + \left\{K_2 A_{\alpha 1}^{(i_{REF}\, i_\alpha - t)}\right\}^2} \\ &\times \cos\left(\tan^{-1}\left(\frac{K_1 A_{\alpha 1}^{(v_\alpha - v_{REF})}}{K_2 A_{\alpha 1}^{(v_{REF}\, v_\alpha - t)}}\right) - \tan^{-1}\left(\frac{K_1 A_{\alpha 1}^{(i_\alpha - i_{REF})}}{K_2 A_{\alpha 1}^{(i_{REF}\, i_\alpha - t)}}\right)\right) \end{aligned} \tag{16.69}$$

Amplitude and phase angle of fundamental component of voltage along β axis in Clarke Plane is obtained from (15.20) and (15.21) as

$$V_{\beta 1} = \sqrt{\left\{K_1 A_{\beta 1}^{(v_\beta - v_{REF})}\right\}^2 + \left\{K_2 A_{\beta 1}^{(v_{REF}\, v_\beta - t)}\right\}^2} \tag{16.70}$$

$$\varphi_{\beta 1} = \tan^{-1}\left(\frac{K_1 A_{\beta 1}^{(v_\beta - v_{REF})}}{K_2 A_{\beta 1}^{(v_{REF}\, v_\beta - t)}}\right) \tag{16.71}$$

Amplitude and phase angle of fundamental component of current along β axis in Clarke Plane is obtained from (15.26) and (15.27) as

$$I_{\beta 1} = \sqrt{\left\{K_1 A_{\beta 1}^{(i_\beta - i_{REF})}\right\}^2 + \left\{K_2 A_{\beta 1}^{(i_{REF}\, i_\beta - t)}\right\}^2} \tag{16.72}$$

$$\theta_{\beta 1} = \tan^{-1}\left(\frac{K_1 A_{\beta 1}^{(i_\beta - i_{REF})}}{K_2 A_{\beta 1}^{(i_{REF}\, i_\beta - t)}}\right) \tag{16.73}$$

Using (16.70) to (16.73), reactive and active power contributed by fundamental components along β axis can be expressed as

$$\begin{aligned} Q_{\beta 1} &= \frac{1}{2} V_{\beta 1} I_{\beta 1} \sin(\varphi_{\beta 1} - \theta_{\beta 1}) \\ &= \frac{1}{2}\sqrt{\left\{K_1 A_{\beta 1}^{(v_\beta - v_{REF})}\right\}^2 + \left\{K_2 A_{\beta 1}^{(v_{REF}\, v_\beta - t)}\right\}^2} \\ &\times \sqrt{\left\{K_1 A_{\beta 1}^{(i_\beta - i_{REF})}\right\}^2 + \left\{K_2 A_{\beta 1}^{(i_{REF}\, i_\beta - t)}\right\}^2} \\ &\times \sin\left(\tan^{-1}\left(\frac{K_1 A_{\beta 1}^{(v_\beta - v_{REF})}}{K_2 A_{\beta 1}^{(v_{REF}\, v_\beta - t)}}\right) - \tan^{-1}\left(\frac{K_1 A_{\beta 1}^{(i_\beta - i_{REF})}}{K_2 A_{\beta 1}^{(i_{REF}\, i_\beta - t)}}\right)\right) \end{aligned} \tag{16.74}$$

$$\begin{aligned} P_{\beta 1} &= \frac{1}{2} V_{\beta 1} I_{\beta 1} \cos(\varphi_{\beta 1} - \theta_{\beta 1}) \\ &= \frac{1}{2}\sqrt{\left\{K_1 A_{\beta 1}^{(v_\beta - v_{REF})}\right\}^2 + \left\{K_2 A_{\beta 1}^{(v_{REF}\, v_\beta - t)}\right\}^2} \\ &\times \sqrt{\left\{K_1 A_{\beta 1}^{(i_\beta - i_{REF})}\right\}^2 + \left\{K_2 A_{\beta 1}^{(i_{REF}\, i_\beta - t)}\right\}^2} \\ &\times \cos\left(\tan^{-1}\left(\frac{K_1 A_{\beta 1}^{(v_\beta - v_{REF})}}{K_2 A_{\beta 1}^{(v_{REF}\, v_\beta - t)}}\right) - \tan^{-1}\left(\frac{K_1 A_{\beta 1}^{(i_\beta - i_{REF})}}{K_2 A_{\beta 1}^{(i_{REF}\, i_\beta - t)}}\right)\right) \end{aligned} \tag{16.75}$$

Total power contributed by fundamental components of voltage and current in Clarke Plane [1], can be written as

$$S_{C1} = (P_{\alpha 1} + P_{\beta 1}) + j(Q_{\alpha 1} + Q_{\beta 1}) \tag{16.76}$$

16.7.2 Contribution of Harmonic Components

Amplitude and phase angle of harmonic component of mth order of voltage along α axis in Clarke Plane is obtained from (15.28) and (15.29) as

$$V_{\alpha m} = \sqrt{\left\{K_1 A_{\alpha m}^{(v_\alpha - v_{REF})}\right\}^2 + \left\{K_2 A_{\alpha m}^{(v_{REF}\, v_\alpha - t)}\right\}^2} \tag{16.77}$$

$$\varphi_{\alpha m} = \tan^{-1}\left(\frac{K_1 A_{\alpha m}^{(v_\alpha - v_{REF})}}{K_2 A_{\alpha m}^{(v_{REF}\, v_\alpha - t)}}\right) \tag{16.78}$$

Amplitude and phase angle of harmonic component of nth order of current along α axis in Clarke Plane is obtained from (15.32) and (15.33) as

$$I_{\alpha n} = \sqrt{\left\{K_1 A_{\alpha n}^{(i_\alpha - i_{REF})}\right\}^2 + \left\{K_2 A_{\alpha n}^{(i_{REF}\, i_\alpha - t)}\right\}^2} \tag{16.79}$$

$$\theta_{\alpha n} = \tan^{-1}\left(\frac{K_1 A_{\alpha n}^{(i_\alpha - i_{REF})}}{K_2 A_{\alpha n}^{(i_{REF}\, i_\alpha - t)}}\right) \tag{16.80}$$

Using (16.77) to (16.80), reactive and active power contributed by harmonic components along α axis can be expressed as

$$\begin{aligned} Q_{\alpha n} &= \frac{1}{2} V_{\alpha n} I_{\alpha n} \sin(\varphi_{\alpha n} - \theta_{\alpha n}) \\ &= \frac{1}{2}\sqrt{\left\{K_1 A_{\alpha n}^{(v_\alpha - v_{REF})}\right\}^2 + \left\{K_2 A_{\alpha n}^{(v_{REF}\, v_\alpha - t)}\right\}^2} \\ &\quad \times \sqrt{\left\{K_1 A_{\alpha n}^{(i_\alpha - i_{REF})}\right\}^2 + \left\{K_2 A_{\alpha n}^{(i_{REF}\, i_\alpha - t)}\right\}^2} \\ &\quad \times \sin\left(\tan^{-1}\left(\frac{K_1 A_{\alpha n}^{(v_R - v_{REF})}}{K_2 A_{\alpha n}^{(v_{REF}\, v_\alpha - t)}}\right) - \tan^{-1}\left(\frac{K_1 A_{\alpha n}^{(i_\alpha - i_{REF})}}{K_2 A_{\alpha n}^{(i_{REF}\, i_\alpha - t)}}\right)\right) \end{aligned} \tag{16.81}$$

$$\begin{aligned} P_{\alpha n} &= \frac{1}{2} V_{\alpha n} I_{\alpha n} \cos(\varphi_{\alpha n} - \theta_{\alpha n}) \\ &= \frac{1}{2}\sqrt{\left\{K_1 A_{\alpha n}^{(v_\alpha - v_{REF})}\right\}^2 + \left\{K_2 A_{\alpha n}^{(v_{REF}\, v_\alpha - t)}\right\}^2} \\ &\quad \times \sqrt{\left\{K_1 A_{\alpha n}^{(i_\alpha - i_{REF})}\right\}^2 + \left\{K_2 A_{\alpha n}^{(i_{REF}\, i_\alpha - t)}\right\}^2} \\ &\quad \times \cos\left(\tan^{-1}\left(\frac{K_1 A_{\alpha n}^{(v_\alpha - v_{REF})}}{K_2 A_{\alpha n}^{(v_{REF}\, v_\alpha - t)}}\right) - \tan^{-1}\left(\frac{K_1 A_{\alpha n}^{(i_\alpha - i_{REF})}}{K_2 A_{\alpha n}^{(i_{REF}\, i_\alpha - t)}}\right)\right) \end{aligned} \tag{16.82}$$

Amplitude and phase angle of mth order harmonic component of voltage along β axis in Clarke plane is obtained from (15.38) and (15.39) as

$$V_{\beta m} = \sqrt{\left\{K_1 A_{\beta m}^{(v_\beta - v_{REF})}\right\}^2 + \left\{K_2 A_{\beta m}^{(v_{REF}\, v_\beta - t)}\right\}^2} \tag{16.83}$$

$$\varphi_{\beta m} = \tan^{-1}\left(\frac{K_1 A_{\beta m}^{(v_\beta - v_{REF})}}{K_2 A_{\beta m}^{(v_{REF}\, v_\beta - t)}}\right) \tag{16.84}$$

Amplitude and phase angle of nth order harmonic component of current along β axis in Clarke plane is obtained from (15.42) and (15.43) as

$$I_{\beta n} = \sqrt{\left\{K_1 A_{\beta n}^{(i_\beta - i_{REF})}\right\}^2 + \left\{K_2 A_{\beta n}^{(i_{REF}\, i_\beta - t)}\right\}^2} \tag{16.85}$$

$$\theta_{\beta n} = \tan^{-1}\left(\frac{K_1 A_{\beta n}^{(i_\alpha - i_{REF})}}{K_2 A_{\beta n}^{(i_{REF}\, i_\beta - t)}}\right) \tag{16.86}$$

Using (16.83) to (16.86), reactive and active power contributed by harmonic components along β axis can be expressed as

$$\begin{aligned}
Q_{\beta n} &= \frac{1}{2} V_{\beta n} I_{\beta n} \sin(\varphi_{\beta n} - \theta_{\beta n}) \\
&= \frac{1}{2}\sqrt{\left\{K_1 A_{\beta n}^{(v_\beta - v_{REF})}\right\}^2 + \left\{K_2 A_{\beta n}^{(v_{REF}\, v_\beta - t)}\right\}^2} \\
&\times \sqrt{\left\{K_1 A_{\beta n}^{(i_\beta - i_{REF})}\right\}^2 + \left\{K_2 A_{\beta n}^{(i_{REF}\, i_{\beta\alpha} - t)}\right\}^2} \\
&\times \sin\left(\tan^{-1}\left(\frac{K_1 A_{\beta n}^{(v_\beta - v_{REF})}}{K_2 A_{\beta n}^{(v_{REF}\, v_\beta - t)}}\right) - \tan^{-1}\left(\frac{K_1 A_{\beta n}^{(i_\beta - i_{REF})}}{K_2 A_{\beta n}^{(i_{REF}\, i_\beta - t)}}\right)\right)
\end{aligned} \tag{16.87}$$

$$\begin{aligned}
P_{\beta n} &= \frac{1}{2} V_{\beta n} I_{\beta n} \cos\left(\varphi_{\beta n} - \theta_{\beta n}\right) \\
&= \frac{1}{2}\sqrt{\left\{K_1 A_{\beta n}^{(v_\beta - v_{REF})}\right\}^2 + \left\{K_2 A_{\beta n}^{(v_{REF}\, v_\beta - t)}\right\}^2} \\
&\times \sqrt{\left\{K_1 A_{\beta n}^{(i_\beta - i_{REF})}\right\}^2 + \left\{K_2 A_{\beta n}^{(i_{REF}\, i_\beta - t)}\right\}^2} \\
&\times \cos\left(\tan^{-1}\left(\frac{K_1 A_{\beta n}^{(v_\beta - v_{REF})}}{K_2 A_{\beta n}^{(v_{REF}\, v_\beta - t)}}\right) - \tan^{-1}\left(\frac{K_1 A_{\beta n}^{(i_\beta - i_{REF})}}{K_2 A_{\beta n}^{(i_{REF}\, i_\beta - t)}}\right)\right)
\end{aligned} \tag{16.88}$$

Total power contributed by harmonic components of voltage and current in Clarke Plane [1], can be written as

$$S_{Cn} = \sum_{n=2,3,\ldots\ldots\ldots\ldots\ldots} \left\{\left(P_{\alpha n} + P_{\beta n}\right) + j(Q_{\alpha n} + Q_{\beta n})\right\} \tag{16.89}$$

Total power contributed by voltage and current in Clarke Plane [1], can be written as

$$S_C = \sum_{n=1,2,3,\ldots\ldots\ldots\ldots\ldots} \left\{\left(P_{\alpha n} + P_{\beta n}\right) + j(Q_{\alpha n} + Q_{\beta n})\right\} \tag{16.90}$$

16.8 Power Components in Park Plane

16.8.1 Contribution of Fundamental Components

Amplitude and phase angle of fundamental component of voltage along d-axis in Park plane is obtained from (15.59) and (15.60) as

$$V_{d1} = \sqrt{\left\{K_1 A_{d1}^{(v_d - v_{REF})}\right\}^2 + \left\{K_2 A_{d1}^{(v_{REF}\, v_d - t)}\right\}^2} \tag{16.91}$$

$$\varphi_{d1} = \tan^{-1}\left(\frac{K_1 A_{d1}^{(v_d - v_{REF})}}{K_2 A_{d1}^{(v_{REF}\, v_d - t)}}\right) \tag{16.92}$$

Amplitude and phase angle of fundamental component of current along d-axis in Park plane is obtained from (15.63) and (15.64) as

$$I_{d1} = \sqrt{\left\{K_1 A_{d1}^{(i_d - i_{REF})}\right\}^2 + \left\{K_2 A_{d1}^{(i_{REF}\, i_d - t)}\right\}^2} \tag{16.93}$$

$$\theta_{d1} = \tan^{-1}\left(\frac{K_1 A_{d1}^{(i_d - i_{REF})}}{K_2 A_{d1}^{(i_{REF}\, i_d - t)}}\right) \tag{16.94}$$

Using (16.91) to (16.94), reactive and active power contributed by fundamental components along d-axis

$$\begin{aligned} Q_{d1} &= \frac{1}{2} V_{d1} I_{d1} \sin\left(\varphi_{d1} - \theta_{d1}\right) \\ &= \frac{1}{2}\sqrt{\left\{K_1 A_{d1}^{(v_d - v_{REF})}\right\}^2 + \left\{K_2 A_{d1}^{(v_{REF}\, v_d - t)}\right\}^2} \\ &\times \sqrt{\left\{K_1 A_{d1}^{(i_d - i_{REF})}\right\}^2 + \left\{K_2 A_{d1}^{(i_{REF}\, i_d - t)}\right\}^2} \\ &\times \sin\left(\tan^{-1}\left(\frac{K_1 A_{d1}^{(v_d - v_{REF})}}{K_2 A_{d1}^{(v_{REF}\, v_d - t)}}\right) - \tan^{-1}\left(\frac{K_1 A_{d1}^{(i_d - i_{REF})}}{K_2 A_{d1}^{(i_{REF}\, i_d - t)}}\right)\right) \end{aligned} \tag{16.95}$$

$$\begin{aligned} P_{d1} &= \frac{1}{2} V_{d1} I_{d1} \cos\left(\varphi_{d1} - \theta_{d1}\right) \\ &= \frac{1}{2}\sqrt{\left\{K_1 A_{d1}^{(v_d - v_{REF})}\right\}^2 + \left\{K_2 A_{d1}^{(v_{REF}\, v_d - t)}\right\}^2} \\ &\times \sqrt{\left\{K_1 A_{d1}^{(i_d - i_{REF})}\right\}^2 + \left\{K_2 A_{d1}^{(i_{REF}\, i_d - t)}\right\}^2} \\ &\times \cos\left(\tan^{-1}\left(\frac{K_1 A_{d1}^{(v_d - v_{REF})}}{K_2 A_{d1}^{(v_{REF}\, v_d - t)}}\right) - \tan^{-1}\left(\frac{K_1 A_{d1}^{(i_d - i_{REF})}}{K_2 A_{d1}^{(i_{REF}\, i_d - t)}}\right)\right) \end{aligned} \tag{16.96}$$

Amplitude and phase angles of fundamental component of voltage along q-axis in Park plane is obtained from (15.65) and (15.66) as

$$V_{q1} = \sqrt{\left\{K_1 A_{q1}^{(v_q - v_{REF})}\right\}^2 + \left\{K_2 A_{q1}^{(v_{REF}\, v_q - t)}\right\}^2} \tag{16.97}$$

$$\varphi_{q1} = \tan^{-1}\left(\frac{K_1 A_{q1}^{(v_q - v_{REF})}}{K_2 A_{q1}^{(v_{REF}\, v_q - t)}}\right) \tag{16.98}$$

Amplitude and phase angles of fundamental component of current along q-axis in Park plane is obtained from (15.71) and (15.72) as

$$I_{q1} = \sqrt{\left\{K_1 A_{q1}^{(i_q - i_{REF})}\right\}^2 + \left\{K_2 A_{q1}^{(i_{REF}\, i_q - t)}\right\}^2} \tag{16.99}$$

$$\theta_{q1} = \tan^{-1}\left(\frac{K_1 A_{q1}^{(i_q - i_{REF})}}{K_2 A_{q1}^{(i_{REF}\, i_q - t)}}\right) \tag{16.100}$$

Using (16.97) to (16.100), reactive and active power contributed by fundamental components along q-axis

$$\begin{aligned} Q_{q1} &= \frac{1}{2} V_{q1} I_{q1} \sin(\varphi_{q1} - \theta_{q1}) \\ &= \frac{1}{2}\sqrt{\left\{K_1 A_{q1}^{(v_q - v_{REF})}\right\}^2 + \left\{K_2 A_{q1}^{(v_{REF}\, v_q - t)}\right\}^2} \\ &\times \sqrt{\left\{K_1 A_{q1}^{(i_q - i_{REF})}\right\}^2 + \left\{K_2 A_{q1}^{(i_{REF}\, i_q - t)}\right\}^2} \\ &\times \sin\left(\tan^{-1}\left(\frac{K_1 A_{q1}^{(v_q - v_{REF})}}{K_2 A_{q1}^{(v_{REF}\, v_q - t)}}\right) - \tan^{-1}\left(\frac{K_1 A_{q1}^{(i_q - i_{REF})}}{K_2 A_{q1}^{(i_{REF}\, i_q - t)}}\right)\right) \end{aligned} \tag{16.101}$$

$$\begin{aligned} P_{q1} &= \frac{1}{2} V_{q1} I_{q1} \cos(\varphi_{q1} - \theta_{q1}) \\ &= \frac{1}{2}\sqrt{\left\{K_1 A_{q1}^{(v_q - v_{REF})}\right\}^2 + \left\{K_2 A_{q1}^{(v_{REF}\, v_q - t)}\right\}^2} \\ &\times \sqrt{\left\{K_1 A_{q1}^{(i_q - i_{REF})}\right\}^2 + \left\{K_2 A_{q1}^{(i_{REF}\, i_q - t)}\right\}^2} \\ &\times \cos\left(\tan^{-1}\left(\frac{K_1 A_{q1}^{(v_q - v_{REF})}}{K_2 A_{q1}^{(v_{REF}\, v_q - t)}}\right) - \tan^{-1}\left(\frac{K_1 A_{q1}^{(i_q - i_{REF})}}{K_2 A_{q1}^{(i_{REF}\, i_q - t)}}\right)\right) \end{aligned} \tag{16.102}$$

Total power contributed by fundamental components of voltage and current in Park Plane, can be written as

$$S_{P1} = (P_{d1} + P_{q1}) + j(Q_{d1} + Q_{q1}) \tag{16.103}$$

16.8.2 Contribution of Harmonic Components in Park Plane

Amplitude and phase angles of mth order harmonic component of voltage along d-axis in Park plane [3] is obtained from (15.75) and (15.76) as

$$V_{dm} = \sqrt{\left\{K_1 A_{dm}^{(v_d - v_{REF})}\right\}^2 + \left\{K_2 A_{dm}^{(v_{REF}\, v_d - t)}\right\}^2} \tag{16.104}$$

$$\varphi_{dm} = \tan^{-1}\left(\frac{K_1 A_{dm}^{(v_d - v_{REF})}}{K_2 A_{dm}^{(v_{REF}\, v_d - t)}}\right) \tag{16.105}$$

Amplitude and phase angles of nth order harmonic component of current along d-axis in Park plane is obtained from (15.79) and (15.80) as

$$I_{dn} = \sqrt{\left\{K_1 A_{dn}^{(i_d - i_{REF})}\right\}^2 + \left\{K_2 A_{dn}^{(i_{REF}\, i_d - t)}\right\}^2} \tag{16.106}$$

$$\theta_{dn} = \tan^{-1}\left(\frac{K_1 A_{dn}^{(i_d - i_{REF})}}{K_2 A_{dn}^{(i_{REF}\, i_d - t)}}\right) \tag{16.107}$$

Using (16.104) to (16.107), reactive and active power contributed by harmonic components along d-axis

$$\begin{aligned} Q_{dn} &= \frac{1}{2} V_{dn} I_{dn} \sin(\varphi_{dn} - \theta_{dn}) \\ &= \frac{1}{2}\sqrt{\left\{K_1 A_{dn}^{(v_d - v_{REF})}\right\}^2 + \left\{K_2 A_{dn}^{(v_{REF}\, v_d - t)}\right\}^2} \\ &\quad \times \sqrt{\left\{K_1 A_{dn}^{(i_d - i_{REF})}\right\}^2 + \left\{K_2 A_{dn}^{(i_{REF}\, i_d - t)}\right\}^2} \\ &\quad \times \sin\left(\tan^{-1}\left(\frac{K_1 A_{dn}^{(v_d - v_{REF})}}{K_2 A_{dn}^{(v_{REF}\, v_d - t)}}\right) - \tan^{-1}\left(\frac{K_1 A_{dn}^{(i_d - i_{REF})}}{K_2 A_{dn}^{(i_{REF}\, i_d - t)}}\right)\right) \end{aligned} \tag{16.108}$$

$$\begin{aligned} P_{dn} &= \frac{1}{2} V_{dn} I_{dn} \cos(\varphi_{dn} - \theta_{dn}) \\ &= \frac{1}{2}\sqrt{\left\{K_1 A_{dn}^{(v_d - v_{REF})}\right\}^2 + \left\{K_2 A_{dn}^{(v_{REF}\, v_d - t)}\right\}^2} \end{aligned}$$

$$\times \sqrt{\left\{K_1 A_{dn}^{(i_d - i_{REF})}\right\}^2 + \left\{K_2 A_{dn}^{(i_{REF}\, i_d - t)}\right\}^2}$$
$$\times \cos\left(\tan^{-1}\left(\frac{K_1 A_{dn}^{(v_d - v_{REF})}}{K_2 A_{dn}^{(v_{REF}\, v_d - t)}}\right) - \tan^{-1}\left(\frac{K_1 A_{dn}^{(i_d - i_{REF})}}{K_2 A_{dn}^{(i_{REF}\, i_d - t)}}\right)\right) \tag{16.109}$$

Amplitude and phase angles of mth order harmonic component of voltage along q-axis in Park plane is obtained from (15.83) and (15.84) as

$$V_{qm} = \sqrt{\left\{K_1 A_{qm}^{(v_q - v_{REF})}\right\}^2 + \left\{K_2 A_{qm}^{(v_{REF}\, v_q - t)}\right\}^2} \tag{16.110}$$

$$\varphi_{qm} = \tan^{-1}\left(\frac{K_1 A_{qm}^{(v_q - v_{REF})}}{K_2 A_{qm}^{(v_{REF}\, v_q - t)}}\right) \tag{16.111}$$

Amplitude and phase angles of nth order harmonic component of current along q-axis in Park plane is obtained from (15.87) and (15.88) as

$$I_{qn} = \sqrt{\left\{K_1 A_{qn}^{(i_q - i_{REF})}\right\}^2 + \left\{K_2 A_{qn}^{(i_{REF}\, i_q - t)}\right\}^2} \tag{16.112}$$

$$\theta_{qn} = \tan^{-1}\left(\frac{K_1 A_{qn}^{(i_q - i_{REF})}}{K_2 A_{qn}^{(i_{REF}\, i_q - t)}}\right) \tag{16.113}$$

Using (16.110) to (16.113), reactive and active power contributed by harmonic components along q-axis

$$Q_{qn} = \frac{1}{2} V_{qn} I_{qn} \sin(\varphi_{qn} - \theta_{qn})$$
$$= \frac{1}{2}\sqrt{\left\{K_1 A_{qn}^{(v_q - v_{REF})}\right\}^2 + \left\{K_2 A_{qn}^{(v_{REF}\, v_q - t)}\right\}^2}$$
$$\times \sqrt{\left\{K_1 A_{qn}^{(i_q - i_{REF})}\right\}^2 + \left\{K_2 A_{qn}^{(i_{REF}\, i_q - t)}\right\}^2}$$
$$\times \sin\left(\tan^{-1}\left(\frac{K_1 A_{qn}^{(v_q - v_{REF})}}{K_2 A_{qn}^{(v_{REF}\, v_q - t)}}\right) - \tan^{-1}\left(\frac{K_1 A_{qn}^{(i_q - i_{REF})}}{K_2 A_{qn}^{(i_{REF}\, i_q - t)}}\right)\right) \tag{16.114}$$

$$P_{qn} = \frac{1}{2} V_{qn} I_{qn} \cos(\phi_{qn} - \theta_{qn})$$
$$= \frac{1}{2}\sqrt{\left\{K_1 A_{qn}^{(v_q - v_{REF})}\right\}^2 + \left\{K_2 A_{qn}^{(v_{REF}\, v_q - t)}\right\}^2}$$
$$\times \sqrt{\left\{K_1 A_{qn}^{(i_q - i_{REF})}\right\}^2 + \left\{K_2 A_{qn}^{(i_{REF}\, i_q - t)}\right\}^2}$$

$$\times\ \cos\left(\tan^{-1}\left(\frac{K_1 A_{qn}^{(v_q - v_{REF})}}{K_2 A_{qn}^{(v_{REF}\, v_q - t)}}\right) - \tan^{-1}\left(\frac{K_1 A_{qn}^{(i_q - i_{REF})}}{K_2 A_{qn}^{(i_{REF}\, i_q - t)}}\right)\right) \tag{16.115}$$

Total power contributed by harmonic components of voltage and current in Park Plane [4], can be written as

$$S_{Pn} = \sum_{n=2,3,\ldots\ldots\ldots\ldots\ldots} \left\{ \left(P_{dn} + P_{qn}\right) + j\left(Q_{dn} + Q_{qn}\right) \right\} \tag{16.116}$$

Total power contributed by fundamental and harmonics of voltage and current in Park Plane, can be written as

$$S_P = \sum_{n=1,2,3,\ldots\ldots\ldots\ldots\ldots} \left\{ \left(P_{dn} + P_{qn}\right) + j\left(Q_{dn} + Q_{qn}\right) \right\} \tag{16.117}$$

16.9 CMS Equations for Power Distortion Factors

16.9.1 Active Power Distortion Factor in Phase R

Active power distortion factor in phase R is given by

$$PDF_R = \frac{\sum\limits_{n=2,3,\ldots\ldots\ldots} \frac{1}{2} V_{Rn} I_{Rn} \cos\left(\phi_{Rn}\right)}{\frac{1}{2} V_{R1} I_{R1} \cos\phi_{R1}} = \frac{\sum\limits_{n=2,3,\ldots.} P_{Rn}}{P_1} \tag{16.118}$$

16.9.2 Reactive Power Distortion Factor in Phase R

Reactive power distortion factor in phase R is given by

$$QDF_R = \frac{\sum\limits_{n=2,3,\ldots\ldots\ldots} \frac{1}{2} V_{Rn} I_{Rn} \sin\left(\phi_{Rn}\right)}{\frac{1}{2} V_{R1} I_{R1} \sin\phi_{R1}} = \frac{\sum\limits_{n=2,3,\ldots.} Q_{Rn}}{Q_1} \tag{16.119}$$

16.9.3 Apparent Power Distortion Factor in Phase R

$$ADF_R = \frac{\sum\limits_{n=2,3,\ldots\ldots\ldots} V_{Rn} I_{Rn}}{V_{R1} I_{R1}} \tag{16.120}$$

Similarly Y and B phases power distortion factors can easily be obtained.

16.9.4 Active Power Distortion Factor in Clarke Plane

Active power distortion factor (PDF) i.e., the ratio of active power contributed by harmonic components to the active power contributed by fundamental component is written as

$$PDF_C = \frac{P_H}{P_1} = \frac{\sum\limits_{n=2,3,\ldots} P_n}{P_1} = \frac{\sum\limits_{n=2,3,\ldots} P_{\alpha n} + \sum\limits_{n=2,3,\ldots} P_{\beta n}}{P_{\alpha 1} + P_{\beta 1}} \tag{16.121}$$

16.9.5 Reactive Power Distortion Factor in Clarke Plane

Reactive power distortion factor (QDF) i.e., the ratio of reactive power contributed by harmonic components to the reactive power contributed by fundamental component is written as

$$QDF_C = \frac{Q_H}{Q_1} = \frac{\sum\limits_{n=2,3,\ldots} Q_n}{Q_1} = \frac{\sum\limits_{n=2,3,\ldots} Q_{\alpha n} + \sum\limits_{n=2,3,\ldots} Q_{\beta n}}{Q_{\alpha 1} + Q_{\beta 1}} \tag{16.122}$$

16.9.6 Active Power Distortion Factor in Park Plane

Real power distortion factor, the ratio of active power contributed by harmonic components to that contributed by fundamental component, is given by

$$PDF_P = \frac{P_H}{P_1} = \frac{\sum\limits_{n=2,3,\ldots} P_n}{P_1} = \frac{\sum\limits_{n=2,3,\ldots} P_{dn} + \sum\limits_{n=2,3,\ldots} P_{qn}}{P_{d1} + P_{q1}} \tag{16.123}$$

16.9.7 Reactive Power Distortion Factor in Park Plane

Reactive power distortion factor, the ratio of reactive power contributed by harmonic components to that contributed by fundamental component, is given by

$$QDF_R = \frac{Q_H}{Q_1} = \frac{\sum\limits_{n=2,3,\ldots} Q_n}{Q_1} = \frac{\sum\limits_{n=2,3,\ldots} Q_{dn} + \sum\limits_{n=2,3,\ldots} Q_{qn}}{Q_{d1} + Q_{q1}} \tag{16.124}$$

16.10 Discussion

The rule set developed in the first part of the chapter is capable of measuring active and reactive power. It can measure power factor and also indicate maximum or minimum reactive or active power. But, main limitation of this method is that the rule set is not capable of giving information of powers contributed by harmonic components. This has been overcome by applying area based approach in power component assessment. For three phase analysis Clarke and Park planes have been used. In computer simulation, fundamental components as well as harmonic components have been assessed wherefrom, different power, i.e., active, reactive and apparent powers have been calculated. Different distortion factors with respect to different power have also been assessed

References

[1] Chattopadhyay, S., Mitra, M., Sengupta, S.: Area based approach for three phase power quality assessment in Clarke plane. J. Electrical Syst. **4**(3), (2008)

[2] Chattopadhyay, S., Mitra, M., Sengupta, S.: Area based mathematical model for harmonic distortion. POWERCON 2008, International Conference on power system organized by IEEE in New Delhi, on 12–15 October 2008

[3] Chattopadhyay, S., Mitra, M., Sengupta, S.: Harmonic analysis in a three-phase system using park transformation technique. Modeling-A, AMSE Int. J. Model. Simul. **80**(3), France (2007)

[4] Chattopadhyay, S., Mitra, M., Sengupta, S.: Single cycle mathematical model for harmonic power. Proceedings of MS'07-India: International Conference on Modeling and Simulation, Kolkata, 3–5 December 2007

[5] Chattopadhyay, S., Mitra, M., Sengupta, S.: Advanced graphics application to determine reactive power and direction of power flow. Proceedings of PEITSICON-2005: International Conference on IT and Power System by IEE (Calcutta section), January 2005

[6] Chattopadhyay, S., Mitra, M., Sengupta, S.: Computerized assessment of electric power quality. Proceedings of 12th State Science Congress organized by DST and Government of West Bengal, Kolkata, February–March 2005

[7] Chattopadhyay, S., Mitra, M., Sengupta, S.: Area based approach for harmonic analysis. Proceedings of 14th State Science Congress organized by DST and Govt. of West Bengal, Kolkata, February–March 2007

[8] Chattopadhyay, S., Mitra, M., Sengupta, S.: Harmonic power in Clarke plane. IE(I). J. Electr. Eng. **91**, 23–28 (2010)

Chapter 17
Transients Analysis

Abstract This chapter describes transients as a power quality problem. Different measurement techniques are discussed followed by their merits and demerits.

17.1 Introduction

Transient analysis is a difficult task because they are unpredictable with respect to the time of occurrence as well as nature of the transients. Moreover transients are events of high peak and short duration [1–3]. Transient analysis deals the identification of transient and its source. For this purpose characterization [1, 4–7] of voltage disturbances is necessary to obtain information of the cause of the disturbances. For oscillatory transients, the amplitude, frequency, and damping of the dominant components are appropriate features. These features are decided by the capacitance and inductance present in the system. For unidirectional transients, peak, rate of rise, rate of fall and duration are important. Research on transients is progressing in different directions. Transient analysis using EMTP is very common [8]. Here, other three important ways for transient analysis are mentioned in the following sections [9, 10, 11].

17.2 Sub-band Filters

Time-frequency analysis is often applied for studying the characteristics of a signal in different frequency bands. Time-evolved signal components in the pre-selected frequency bands are assessed using time frequency analysis. Transients are normally events of short duration. For this reason Short time Fourier transform (STFT) is one useful tool for transient analysis. STFT is a time frequency signal decomposition method. This requires a set of band pass filters with an equal bandwidth. Wavelet filters are another tools for time frequency (or, time scale) analysis of transients. By this method, the signal is decomposed in to time evolved components with

S. Chattopadhyay et al., *Electric Power Quality,* Power Systems,
DOI 10.1007/978-94-007-0635-4_17,

octave bandwidths. These methods are also utilized for the analysis of power system transients.

Limitation: The product of time resolution and frequency resolution remains a constant governed by the uncertainty principle. For this reason, analysis using these sub-band filters suffers from the time frequency resolution constraint. A short time window provides a better time resolution but in a band pass filter with a larger band width and, thus, worse frequency resolution. The optimization between the time and frequency resolution imposes problems in the analysis of signal containing components with closely spaced frequencies. Another limitation of this method is the difficulty to interpret the resulting time varying spectra. Recognition of patterns corresponding to events can be manually possible, but automatic analysis is very difficult.

17.3 Model Based Approaches

Model based approach is very effective for estimating the dominant components of the signal at unknown frequencies, which could be close to each other. High frequency-resolution model based approaches are used in transient analysis to estimate the frequencies and extract the frequency components. The main frequency components in most power system transients are due to the resonance of the system. The parameters of those frequency components carry information on the system and the underlying event.

Most of the transients shows sum of exponentially decaying sinusoids of short duration with additive white noise $e(t_k)$. in some work, they are modeled as

$$x(t_k) = \sum_{i=1}^{p} A_i\, e^{-\alpha_i t_k} \cos(\omega_i t_k + \phi_i) u(t_k - t_i) + e(t_k) \tag{17.1}$$

where $u(t_k - t_i)$ is a unit step function defined as

$$u(t_k - t_i) = \begin{cases} 1, & t_k \geq t_i \\ 0, & t_k < t_i \end{cases} \tag{17.2}$$

Many attempts have been taken to estimate those parameters. Prony's method and ESPRIT are two possible ways of estimating the parameters of such a model. They are used to determine the frequencies, the damping coefficients, the initial phases and the amplitude under the assumption that these parameter do not change within the analysis window. For the analysis of short duration transients in power systems, both methods can be used. Prony's method based on pole-zero models was proposed for the analysis of transients due to single causes for example earth fault currents, whereas ESPRIT method can be used to analyze a range of transients due to different causes.

17.4 ESPRIT Method

Making an assumption that the underlying frequencies are exponentially damped sinusoids, ESPRIT is used to resolve closely spaced frequency components of a signal. By ESPRIT signal is decomposed into a sum of sinusoids using the signal subspace-based approach. For this purpose, prefilering is needed to the voltage waveforms before applying ESPRIT. First, a high pass linear-phase finite-impulse-response (FIR) filter having cut-off frequency 200 Hz (normally) is applied to remove the fundamental frequency component. In most practical cases, the fundamental frequency component is predominant having magnitude 10–100 times larger than the remaining components. This may be the source of large estimation errors for the relatively weaker signal components when using ESPRIT. Pre-filtering is very effective to resolve this problem. As long as the frequencies of transient components under consideration are much higher than the fundamental frequency, the artifacts of pre-filtering are negligible for the transient analysis. For frequency of transient is close to the power system fundamental frequency, a notch filter at the fundamental frequency can be used.

As mentioned before, it is important that the ESPRIT method be applied to a window in which the relevant parameters of the signals are stationary. This requires that the beginning and ending points of the transients are found. After parameter estimation, the transient is reconstructed using the obtained result. The reconstructed signal is used to indicate how well the original transient is modeled and how accurate the parameters are estimated. This can be measured by using mean square error (MSE) criterion between the original and the reconstructed signals. It should be noted that ESPRIT is applied to a part of the transient that starts after all switching actions are completed. Further, it should be noted that all components start at the same time instant; therefore, if a transient under analysis contains frequency elements that start later, then the results of ESPRIT will be reliable.

17.5 Suitability of ESPRIT

Oscillatory transients are modeled by damped sinusoids by ESPRIT. As long as the model fits the signal well, ESPRIT provides good result of estimation for dissolving signal components at closely spaced frequencies with less error. However, for impulsive transients, ESPRIT is not a suitable analysis tool. The major components of transients obtained from ESPRIT provide a means of identifying the possible causes of transients.

Advantage: ESPRIT uses damped sinusoidal signal modeling and is able to provide good estimation of signal components with a very high frequency resolution for signals in noise. Oscillatory transients that obey the damped sinusoidal models, ESPRIT is the best choice among the existing signal processing methods. There is another alternative method called Multiple Signal Classification (MUSIC) that also employs damped sinusoidal model which is based on noise subspace. But reliability MUSIC is still less than other methods.

Non model based approach show relatively low frequency resolution with respect to ESPRIT. So, non-model-based method is not suitable to estimate signal components at arbitrary frequencies. In addition, when the noise is present, the performance is rather poor with non-model based approach. Wavelet is another example of non-model based method that decomposes signal into sub-band components; however, it cannot resolve signal components within the same band.

Limitations: ESPRIT uses the damped sinusoidal model which implies that all signal components start at the same time instant. However, this is not always the case. When a transient is the sum of multiple time delayed transients either with a single or multiple causes, one needs to use a more general damped sinusoidal model. In these cases, ESPRIT has so far not been able to provide a solution to such a case, and the problem remains an open research topic for signal processing research. Researcher can explore wavelets, segmentation, sliding window ESPRIT, etc, to find the starting time instants of different components. Modeling and analysis of impulsive transients remain an open research issue.

17.6 Discussion

Future research on transient analysis includes automatically detecting the start and end of a transient; analyzing multiple transients with a single cause or multiple causes; and extracting information from traveling waves. Others include modeling and analyzing impulsive transients; modeling power system components at high frequencies (e.g., few kilo-hertz); characterizing transients for performance specification of power systems and for understanding their effect on the end user equipment.

References

[1] Bollen, M.H.J., Styvaktakis, E., Yu-HuaGu, I.: Categorization and analysis of power system transients. IEEE Trans. Power Deliv. **20**(3), 2298–2306 (2005)
[2] Arrillaga, J., Watson, N.R., Chen, S.: Power System Quality Assessment. Wiley, New York (2000)
[3] Herath, C., Gosbell, V., Perera, S.: A transient index for reporting power quality (PQ) surveys. In Proceedings CIRED 2003, Bercelona, pp. 2.61-1–2.61-5, Spain, May 2003
[4] Santoso, S., Lamoree, J., Grady, W.M., Powers, E.J., Bhatt, S.C.: A scalable PQ event identification system. IEEE Trans. Power Deliv. **15**, 738–742 (2000)
[5] Santoso, S., Powers, E.J., Grady, W.M., Parsons, A.C.: Characterization of distribution power quality events with fourier and wavelet transform. IEEE Trans. Power Deliv. **15**(1), 247–256 (2000)
[6] Bollen, M.H.J.: Understanding power quality problems. IEEE Press Series on Power Engineering, (2000)

[7] Martzloiff F.D., Gruzs, T.M.: Power quality site surveys: Facts, fiction, and fallacies. IEEE Trans. Ind. Appl. **24**(6), (1998)

[8] Long, W., Cotcher, D., Ruiu, D., Adam, P., Lee, S., Adapa, R.: EMTP—a powerful tool for analyzing power system transients. IEEE Comput. Appl. Power. **3**(3), 36–41 (1990)

[9] Huang, S.-J., Yang, T.-M., Huang, J.-T.: FPGA realization of wavelet transform for detection of electric power system disturbances. IEEE Trans. Power Deliv. **17**(2), 388–394 (2002)

[10] Shaw, S.R., Laughman, C.R., Leeb, S.B., Lepard, R.F.: A power quality prediction system. IEEE Trans. Ind. Electron. **47**(3), (2000)

[11] Simpson, R.H.: Instrumentation, measurement techniques and analytical tools in power quality studies. Proceedings of IEEE, Annual Conference of pulp and paeu industry, pp. 119–140 (1997)

Chapter 18
Passivity and Activity Based Models of Polyphase System

Abstract The chapter develops a passive model of a polyphase system in presence of harmonics. Equivalent circuit and layer based representation of the model have been developed. After describing the main limitation of this model, activity based model has been introduced. Its equivalent circuit and layer based representation have been developed. A case study on induction machine has been done on the basis of activity based model.

18.1 Introduction

A lot of study has been done on modeling of a poly phase system. Some of them have included presence and effects harmonics; harmonics impedance [1] is modeled, distributed networks [2] and multi-converter systems [3] are also included in such modeling. Here modeling of a poly phase system in presence of harmonics is described in a generalized way. In the first part of this chapter, a passivity-based model (PBM) of a poly phase system has been developed; its equivalent circuit and layer-based representation of the passive impedances have been drawn. Discussing its limitations, in the second part of the chapter, an activity-based model (ABM) has been developed in presence harmonics. Its equivalent circuit, control system and layer based representation of active impedances are developed [4]. On the basis of the developed models, a case study has also been made to test the suitability of the model.

18.2 Passivity Based Model

18.2.1 Mathematical Model

Any poly phase system has its own impedance matrix by which voltages and currents are related.

S. Chattopadhyay et al., *Electric Power Quality*, Power Systems,
DOI 10.1007/978-94-007-0635-4_18,

Let,

$[V]$ be the voltage matrix
$[I]$ be the current matrix and
$[Z]$ be the impedance matrix

They can be related as

$$[V] = [Z][I] \tag{18.1}$$

In (18.1), voltage and current have the same frequencies. This implies that, impedance matrix when operating on current matrix, does not change frequency and relates current with voltage having same frequencies. In other words, impedance matrix does not give any information of generation of any new signal having new frequency other than existing system frequencies. Therefore, this impedance can be treated as passive impedance. The impedance matrix $[Z]$ is renamed as passive impedance $[Z_P]$, where the suffix 'p' indicates passive behavior of impedance. Thus (18.1) is rewritten as

$$[V] = [Z_P][I] \tag{18.2}$$

Passive impedance $[Z_P]$ is defined as the impedance which does not give any information of generation of any new signal having new frequency other than existing system frequencies. It relates voltage and current having same frequency.

In (18.2), the frequency of waveforms of voltage matrix is equal to the frequency of waveforms of current matrix and magnitude will depend on the impedance matrix. If the system has only fundamental frequency, the voltage and current matrix will have only fundamental frequency. Thus, for voltage and current with fundamental frequency (18.2) becomes

$$[V_1] = [Z_{p11}][I_1] \tag{18.3}$$

where,

$[V_1]$ = matrix having fundamental frequency voltage
$[I_1]$ = matrix having fundamental frequency current and
$[Z_{P11}]$ = matrix consisting of elements having fundamental frequency impedance

In the same way, if the voltage and current consist of one second order frequency, (18.2) can be written as

$$[V_2] = [Z_{P22}][I_2] \tag{18.4}$$

where, $[V_2]$, $[I_2]$ and $[Z_{P22}]$ indicate voltage, current and impedance matrices respectively corresponding to the frequency of second order harmonics.

In the same way, voltage and current of higher order harmonics can be expressed as

$$[V_3] = [Z_{p33}][I_3]$$

$$[V_4] = [Z_{p_{44}}][I_4]$$

$$\dots\dots\dots\dots\dots\dots$$

$$\dots\dots\dots\dots\dots\dots$$

$$[V_n] = [Z_{p_{nn}}][I_n] \qquad (18.5)$$

where, $[V_n]$, $[I_n]$ and $\left[Z_{p_{nn}}\right]$ indicate voltage, current and impedance matrices respectively corresponding to the frequency of nth order harmonics. (18.3), (18.4) and (18.5) can be expressed by a relation

$$\sum_n [V_n] = \sum_n \left[Z_{p_{nn}}\right][I_n] \qquad (18.6)$$

(18.6) represents all components of voltage and current having same frequency and it consists of passive impedances; hence represents the passive model of a poly phase system [4].

18.2.2 Equivalent Circuit of Passive Model of a Polyphase System

Figure 18.1a represents equivalent circuit of (18.6). It shows that system has impedance matrix $\sum_n [Z_{p_{nn}}]$, voltages applied to the system $\sum_n [V_n]$ and currents

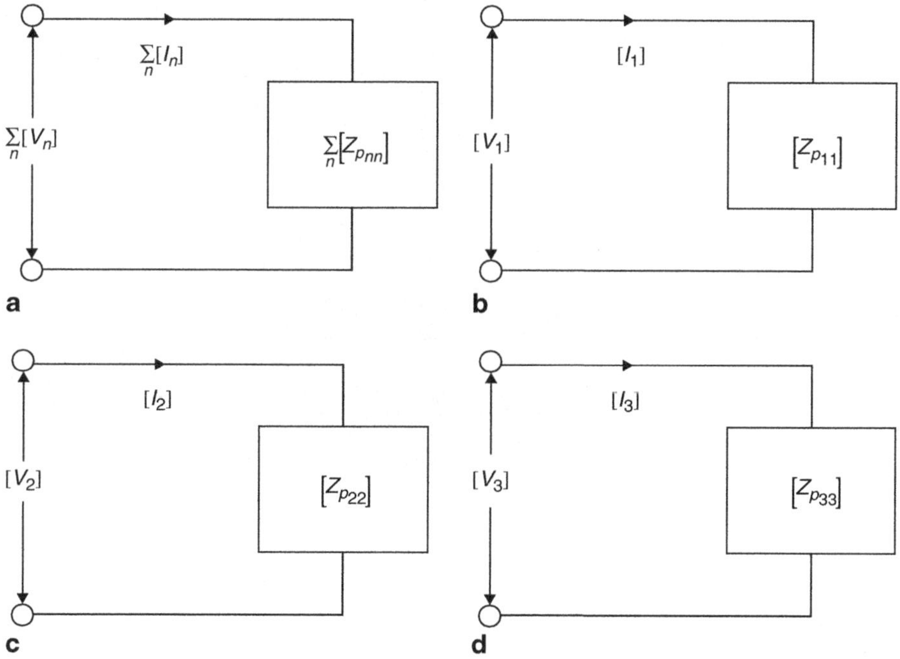

Fig. 18.1 **a** Passive model of a polyphase system. **b** Passive model of a polyphase system for n = 1. **c** Passive model of a poly phase system for n = 2. **d** Passive model of a poly phase system for n = 3

flowing through the system is $\sum_n [I_n]$, where n = 1, 2, 3, ... The impedance of this circuit relates voltage and current of same frequency. The circuit for n = 1, 2 and 3 are shown in Fig. 18.1b, c and d respectively. Each circuit is independent in nature and does not have any mutual interaction among them.

18.2.3 Layer Based Representation of Passive Impedances

Impedances have been represented in voltage-current-frequency plane. Passive impedances relate voltages and currents of same frequency as shown in Fig. 18.1b, c and d. Thus passive impedances lie in the voltage and current plane of same frequency. Figure 18.2 shows the layer based representation of passive impedance.

Horizontal layers of Fig. 18.2 show that each harmonic component has a distinct layer of impedance and there is no interaction between two horizontal layers. Each horizontal layer relates voltage and current matrix of same frequency. For example elements or slopes representing elements of $[Z_{p_{11}}]$ will lie on its layer which is the lowest horizontal layer in the figure. Elements or slopes representing elements of $[Z_{p_{22}}]$ will lie on its layer, which is the second horizontal layer in the figure. In general, $[Z_{p_{nn}}]$ relates voltage and current matrix of nth order of harmonics. This describes the situation where all the harmonic components are supplied from external circuit and no new harmonics are generated inside the system.

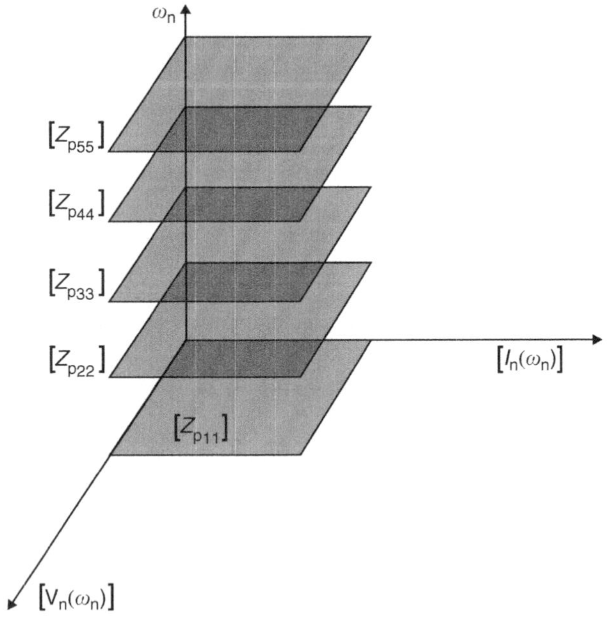

Fig. 18.2 Layer based representation of passive impedance

18.2.4 Limitation of Passive Model

Passive model represented by (18.6) does not give any information about generation of new harmonics inside the system due to interaction of different harmonics present. Since the layers in passive model are parallel to each other, passive model of the system is not capable of giving any information about the harmonics, which may be generated inside the system due to its non-linear behavior [4].

18.3 CMS Activity Based Model

18.3.1 Mathematical Model

The limitation of passive model has been overcome by introducing CMS activity based model. In this model, following factors have been included.

1. Harmonic may be produced inside the system.
2. Highest order of harmonic in voltage may not be equal to the highest order of harmonic in current.
3. Any order of harmonics can produce any other order of harmonics.

Let voltage matrix consists of harmonics up to order ‘m’ and current matrix consists of harmonics up to order ‘n’. Then (18.6) can be rewritten as

$$\sum_{m}[V_m] = \sum_{m,n}[Z_{p_{mn}}][I_n] = \sum_{m \neq n}[Z_{p_{mn}}][I_n] + \sum_{m=n}[Z_{p_{mn}}][I_n] \tag{18.7}$$

In (18.7), $\sum_{m \neq n}[Z_{p_{mn}}]$ represents the impedance due to active sources contributed by interaction among harmonics inside the machine by all causes and it relates mth order voltage matrix with nth order current matrix. $[Z_{p_{nn}}]$ represents passive impedance part of the system which does not generate new harmonics and relates nth order voltage and current matrices. Considering all frequencies, (18.7) can be rewritten as

$$[V_1] = [Z_{p_{11}}][I_1] + [Z_{p_{12}}][I_2] + [Z_{p_{13}}][I_3] + \cdots\cdots [Z_{p_{1n}}][I_n] \tag{18.8}$$

$$[V_2] = [Z_{p_{21}}][I_1] + [Z_{p_{22}}][I_2] + [Z_{p_{23}}][I_3] + \cdots\cdots [Z_{p_{2n}}][I_n] \tag{18.9}$$

$$[V_3] = [Z_{p_{31}}][I_1] + [Z_{p_{32}}][I_2] + [Z_{p_{33}}][I_3] + \cdots\cdots [Z_{p_{3n}}][I_n] \tag{18.10}$$

……

……

$$[V_m] = [Z_{p_{m1}}][I_1] + [Z_{p_{m2}}][I_2] + [Z_{p_{m3}}][I_3] + \cdots\cdots [Z_{p_{mn}}][I_n] \tag{18.11}$$

(18.8) to (18.11) represent voltages of first, second, third... mth order harmonic, which are related to currents of first, second, third,.... nth order harmonic These equations can be written in matrix form as

$$\begin{bmatrix} [V_1] \\ [V_2] \\ [V_3] \\ \cdots \\ \cdots \\ [V_m] \end{bmatrix} = \begin{bmatrix} [Z_{p_{11}}][Z_{p_{12}}][Z_{p_{13}}] & \cdots & \cdots & \cdots & [Z_{p_{1(n-1)}}] & [Z_{p_{1n}}] \\ [Z_{p_{21}}][Z_{p_{22}}][Z_{p_{23}}] & \cdots & \cdots & \cdots & [Z_{p_{2(n-1)}}] & [Z_{p_{2n}}] \\ [Z_{p_{31}}][Z_{p_{32}}][Z_{p_{33}}] & \cdots & \cdots & \cdots & [Z_{p_{3(n-1)}}] & [Z_{p_{3n}}] \\ \cdots\ \cdots\ \cdots & \cdots & \cdots & \cdots & \cdots & \cdots \\ \cdots\ \cdots\ \cdots & \cdots & \cdots & \cdots & \cdots & \cdots \\ [Z_{p_{m1}}][Z_{p_{m2}}][Z_{p_{m3}}] & \cdots & [Z_{pmm}] & \cdots & [Z_{p_{m(n-1)}}] & [Z_{p_{nn}}] \end{bmatrix} \begin{bmatrix} [I_1] \\ [I_2] \\ [I_3] \\ \cdots \\ \cdots \\ [I_n] \end{bmatrix} \tag{18.12}$$

As $[Z_{P_{mn}}]_{m \neq n}$ gives information of interaction among of voltage and current harmonics of different frequencies, it may be resymbolized as $[Z_{a_{mn}}]$ and renamed as active impedance, where suffix 'a' indicates the active nature of the impedance.

Active impedance is defined as the impedance which gives information of generation of new harmonics inside the system and relates voltage and current of different frequencies.

Using active impedances, (18.11) can be written as

$$\begin{bmatrix} [V_1] \\ [V_2] \\ [V_3] \\ \cdots \\ \cdots \\ [V_m] \end{bmatrix} = \begin{bmatrix} [Z_{p_{11}}][Z_{a12}][Z_{a13}] & \cdots & \cdots & \cdots & [Z_{a1(n-1)}] & [Z_{a1n}] \\ [Z_{a21}][Z_{p_{22}}][Z_{a23}] & \cdots & \cdots & \cdots & [Z_{a2(n-1)}] & [Z_{a2n}] \\ [Z_{a31}][Z_{a32}][Z_{p_{33}}] & \cdots & \cdots & \cdots & [Z_{a3(n-1)}] & [Z_{a3n}] \\ \cdots\ \cdots\ \cdots & \cdots & \cdots & \cdots & \cdots & \cdots \\ \cdots\ \cdots\ \cdots & \cdots & \cdots & \cdots & \cdots & \cdots \\ [Z_{am1}][Z_{am2}][Z_{am3}] & \cdots & [Z_{pmm}] & \cdots & [Z_{am(n-1)}] & [Z_{amn}] \end{bmatrix} \begin{bmatrix} [I_1] \\ [I_2] \\ [I_3] \\ \cdots \\ \cdots \\ [I_n] \end{bmatrix} \tag{18.13}$$

where, $[Z_{p_{11}}]$, $[Z_{p_{22}}]$, $[Z_{p_{33}}]\ldots$ are passive impedances and $[Z_{a12}]$, $[Z_{a13}]$, $[Z_{a21}]$, $[Z_{a23}]$, $[Z_{a31}]$, $[Z_{a32}]\ldots$ are active impedances. (18.13) represents an active model of a system consisting of fundamental as well as other harmonic components which may be generated inside the system or supplied from external circuits.

18.3.2 Equivalent Circuit of Active Model

Equivalent circuit of (18.13) is drawn in Fig. 18.3 [4]. It shows that system has impedance matrix $\sum_n [Z_{p_{nn}}]$. Voltages in the system are $\sum_m [V_m]$. Currents flowing through the system are $\sum_n [I_n]$. All the frequencies supplied by from external circuits are present in passive impedance matrix $\sum_n [Z_{p_{nn}}]$. $\sum_{m \neq n} [Z_{a_{mn}}][I_n]$ represents the source or generation of harmonics inside the system, where $[Z_{a_{mn}}]$ is the active impedance generated by mutual interaction of mth order voltage and nth order current.

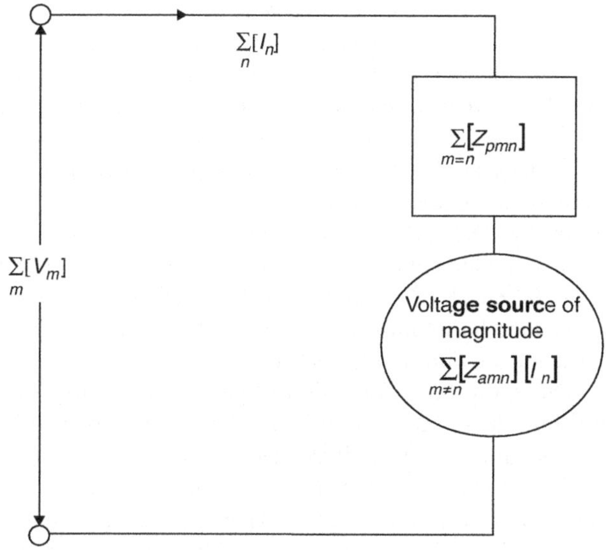

Fig. 18.3 Equivalent circuit of CMS activity based model of a polyphase system

18.3.3 Layer Based Representation of Active Model

Impedance has been represented in voltage-current-frequency plane. Passive impedance relates voltage and currents of same frequency and lies in the voltage and current plane of same frequency. Active impedance relates voltage and currents of different frequency and hence lies in the inclined plane connecting two different frequencies. Figure 18.4 shows the layer based representation of active model of a system having first and second order frequencies.

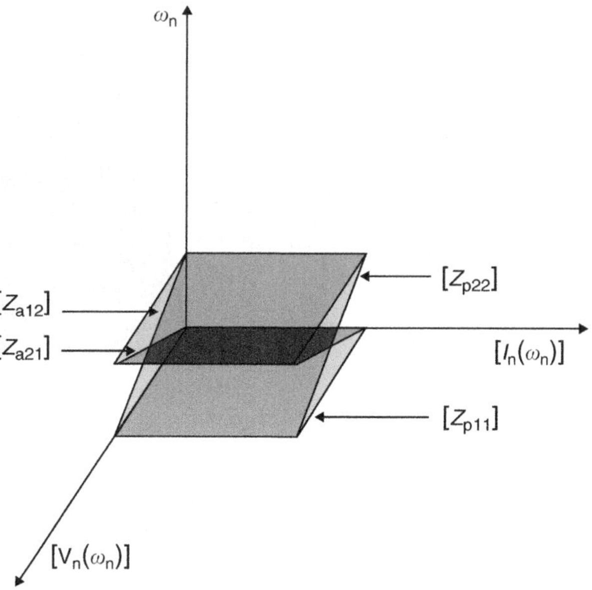

Fig. 18.4 Layer based representation of passive and active impedance

In Fig. 18.4, there are two horizontal layers and two inclined layers. Each horizontal layer relates voltage and current of same frequency. The layer also holds passive impedances. Inclined layers show that each harmonic component is generated from other harmonic component and there is distinct interaction between two layers. Thus inclined planes represent active layers of the system, which describes the situation where the harmonic components are generated inside the system due to mutual interaction of the layers.

In Fig. 18.5, $[Z_{p_{11}}]$is the only passive impedance, which is responsible for producing fundamental current waveform from fundamental supply voltage. $[Za_{61}]$, $[Za_{51}]$, $[Za_{41}]$, $[Za_{31}]$ and $[Za_{21}]$ are active impedances responsible for generation of voltage harmonics of $[V_6\,(\omega_6)]$, $[V_5\,(\omega_5)]$, $[V_4\,(\omega_4)]$, $[V_3\,(\omega_3)]$, $[V_2\,(\omega_2)]$ from fundamental current $[I_1\,(\omega_1)]$ or vice versa.

One more layer formation of passive and active impedance in an active model responsible for generation of fundamental component as well as other harmonics are shown in Fig. 18.6. It includes some more passive and active impedances of the system. The current matrix consists of harmonics up to 6th order which are partly injected by the power supply and partly produced inside the system. Passive layers up to 6th order and have been shown. Also interaction between two consecutive layers and interaction between first layers with all other layers have also been shown.

In all the cases above, harmonics up to 6th order are considered. Figure 18.6 shows the passive impedances up to 6th harmonics, active impedances due to interaction between 1st and 2nd, 2nd and 3rd harmonics. Figures show the passive impedances and active impedances due to interaction among different harmonics.

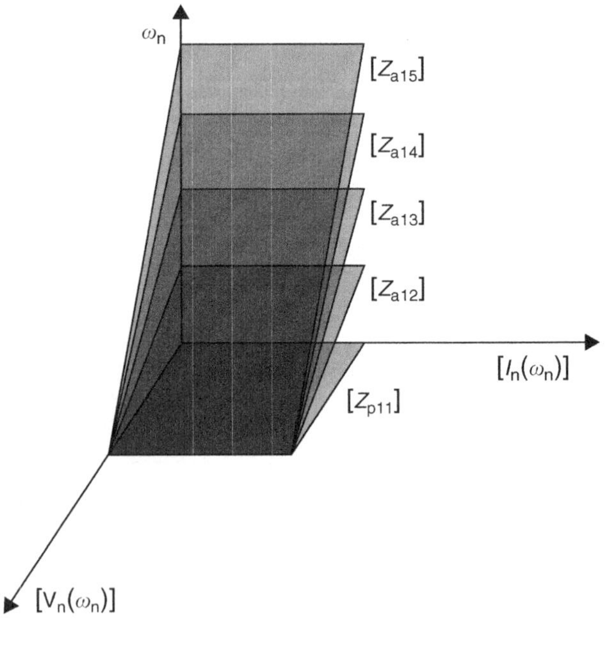

Fig. 18.5 Layer based representation of passive and active impedance responsible for generation of fundamental wave only

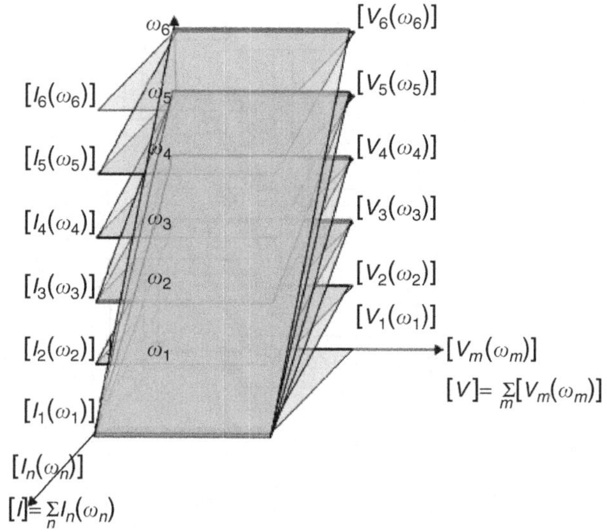

Fig. 18.6 Layer based representation of passive and active impedances

18.4 Mutual Interaction of Voltage and Current of Different Frequencies in Park Plane

Figure 18.7 shows activity based model of a multi-harmonic three-phase system in Park Plane. Supply consists of nth order voltage and current and they are related by passive impedances $[Z_{p_{nn}}]$. Due to the presence of active impedances, voltage of mth order are produced from n^{th} order current by $[Z_{a_{mn}}]$. Voltage of mth order are produced from m^{th} order current by $[Z_{P_{mm}}]^{-1}$. Voltage of nth order will be produced from m^{th} order current by $[Z_{a_{nm}}]$. Current of nth order may flow due to m^{th} order voltage by $[Z_{a_{mn}}]^{-1}$.

18.5 Active Model of a System having Harmonics up to Third Order: A Case Study

For simplicity consider the possibility of harmonics up to third order and in that situation the active model (18.13) can be written as

$$\begin{bmatrix}[V_1]\\ [V_2]\\ [V_3]\end{bmatrix} = \begin{bmatrix}[Z_{p_{11}}][Z_{a12}][Z_{a13}]\\ [Z_{a21}][Z_{p_{22}}][Z_{a23}]\\ [Z_{a31}][Z_{a32}][Z_{p_{33}}]\end{bmatrix}\begin{bmatrix}[I_1]\\ [I_2]\\ [I_3]\end{bmatrix} \tag{18.14}$$

In (18.14), $[Z_{p_{11}}][I_1]$, $[Z_{p_{22}}][I_2]$ and $[Z_{p_{33}}][I_3]$ are the part of voltage matrices which consist of the harmonics supplied by the source. Similarly, $[Z_{a12}][I_2], [Z_{a13}][I_3]$, $[Z_{a21}][I_1], [Z_{a23}][I_3]$ and $[Z_{a31}][I_1], [Z_{a32}][I_2]$ are the

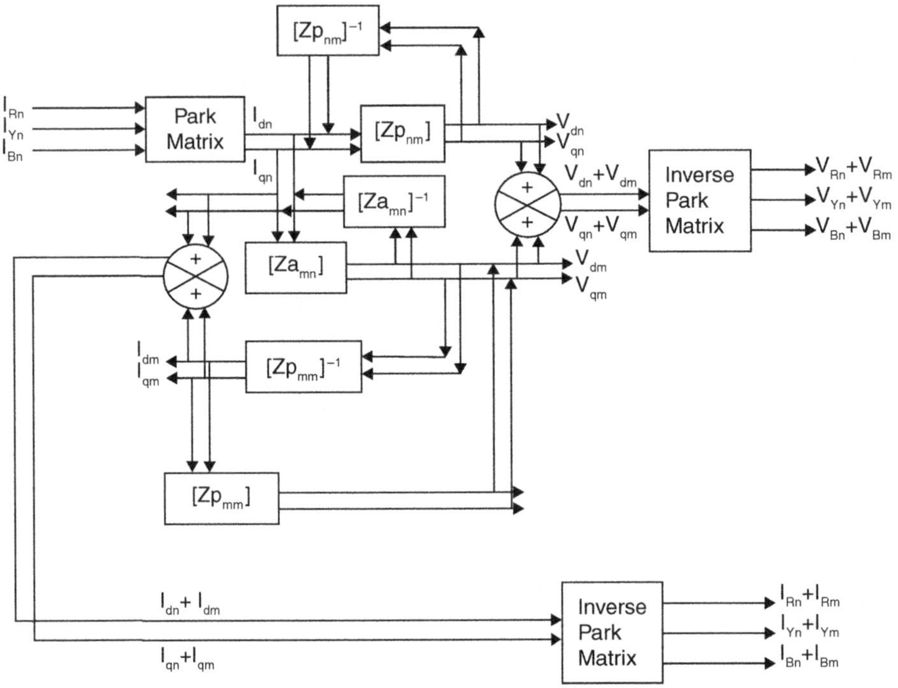

Fig. 18.7 CMS Activity based model of a multi-harmonic three-phase system in Park Plane

part of voltage expressions which consist of harmonics were not supplied by the source but have been generated inside the system.

Now consider there was no other frequency except the fundamental at initial stage. By passive impedance, voltage matrix will contain only fundamental frequency given by

$$[V_1] = \left[Z_{p_{11}}\right][I_1] \tag{18.15}$$

But by the active impedances, harmonics will be created in voltage represented by

$$\begin{bmatrix} [V_1] \\ [V_2] \\ [V_3] \end{bmatrix} = \begin{bmatrix} [Z_{p_{11}}] & [Z_{a12}] & [Z_{a13}] \\ [Z_{a21}] & [Z_{p_{22}}] & [Z_{a23}] \\ [Z_{a31}] & [Z_{a32}] & [Z_{p_{33}}] \end{bmatrix} \begin{bmatrix} [I_1] \\ [0] \\ [0] \end{bmatrix} \tag{18.16}$$

where $[V_2]$ and $[V_3]$ have been created by active components $[Z_{a21}]$ and $[Z_{a31}]$. Now $[V_2]$ and $[V_3]$ will produce current $[I_2]$ and $[I_3]$ given by the inverse of $[Z_{p_{22}}]$ and $[Z_{p_{33}}]$. Then $[I_2]$ and $[I_3]$ will produce other harmonic in the voltage expressions controlled by active impedance $[Z_{a12}]$, $[Z_{a13}]$, $[Z_{a23}]$, $[Z_{a32}]$ and two passive impedance $[Z_{p_{22}}]$ and $[Z_{p_{33}}]$. Then (18.16) becomes,

$$\begin{bmatrix} [V_1] \\ [V_2] \\ [V_3] \end{bmatrix} = \begin{bmatrix} [Z_{p_{11}}] & [Z_{a12}] & [Z_{a13}] \\ [Z_{a21}] & [Z_{p_{22}}] & [Z_{a23}] \\ [Z_{a31}] & [Z_{a32}] & [Z_{p_{33}}] \end{bmatrix} \begin{bmatrix} [I_1] \\ [I_2] \\ [I_3] \end{bmatrix} \tag{18.17}$$

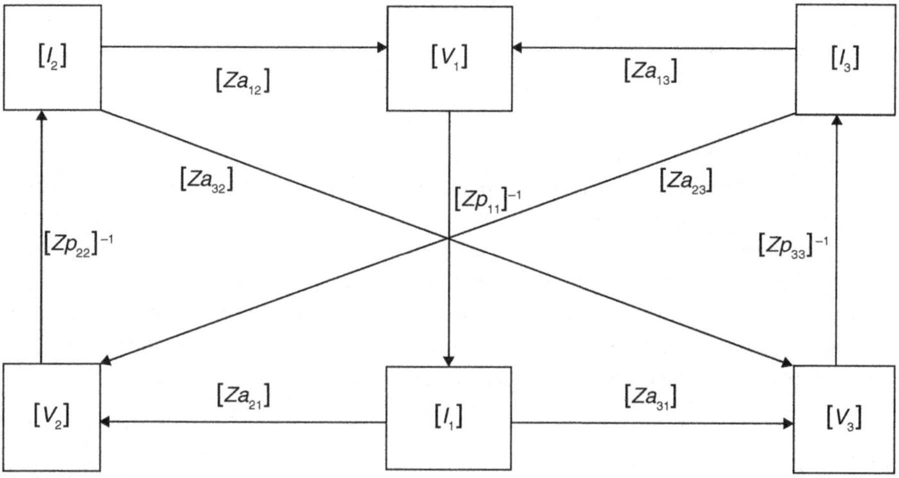

Fig. 18.8 Flow diagram of active model of a system having harmonics up to third order

Thus the generated harmonics depend on the values of active impedances of the circuit. The inter-relations between the harmonic components can be well understood by a flow diagram as shown in Fig. 18.8.

18.6 Nature of Active Impedance

(18.12) is the most general case where harmonics are generated inside the system due to the presence of [Z_{amn}]. Active impedance [Z_{amn}] performs two jobs: cancels the term containing n^{th} order of harmonic and generates m^{th} order of harmonic. Adding with this, [Z_{amn}] controls the amplitude of voltage waveform of frequency of order m. This indicates that [Z_{amn}] should have time dependent function f(t) represented as (18.17) multiplied by time independent function assumed to be equal in form like passive impedance [Z_{pmm}].

$$f(t) = k_{mn} \frac{\sin m\omega t}{\sin n\omega t} \tag{18.18}$$

k_{mn} is constant for a particular system. This constant depends on the design parameter which may vary during a fault.Thus,

$$[Z_{amn}] = k_{mn} \frac{\sin m\omega t}{\sin n\omega t} [Z_{p_{mm}}] \tag{18.19}$$

Thus it seems that active impedance is both time and frequency dependent

18.7 Case Study of Active Model on Poly-phase Induction Machine

A case study has been carried out by developing activity based model of a poly-phase induction machine. First a general activity based model has been considered for a poly-phase induction machine. Then an active model has been developed considering the machine an ideal one, which does not produce harmonics inside the machine. Then a real induction machine has been considered, in which new harmonics are produced and corresponding active model has been developed.

Voltage and current matrix of a rotating machine consist of stator and rotor components which may again be subdivided into d axis and q axis components. Writing these components in voltage and transformed matrix, (18.13) becomes

$$\begin{bmatrix} \begin{bmatrix} V_{qs_1} \\ V_{ds_1} \\ V_{qr_1} \\ V_{dr_1} \end{bmatrix} \\ \begin{bmatrix} V_{qs_2} \\ V_{ds_2} \\ V_{qr_2} \\ V_{dr_2} \end{bmatrix} \\ \begin{bmatrix} V_{qs_3} \\ V_{ds_3} \\ V_{qr_3} \\ V_{dr_3} \end{bmatrix} \\ \cdots \\ \cdots \\ \begin{bmatrix} V_{qs_m} \\ V_{ds_m} \\ V_{qr_m} \\ V_{dr_m} \end{bmatrix} \end{bmatrix} = \begin{bmatrix} [Z_{p_{11}}]\,[Z_{a12}]\,[Z_{a13}] & \cdots & \cdots & \cdots & [Z_{a1(n-1)}] & [Z_{a1n}] \\ [Z_{a21}]\,[Z_{p_{22}}]\,[Z_{a23}] & \cdots & \cdots & \cdots & [Z_{a2(n-1)}] & [Z_{a2n}] \\ [Z_{a31}]\,[Z_{a32}]\,[Z_{p_{33}}] & \cdots & \cdots & \cdots & [Z_{a3(n-1)}] & [Z_{a3n}] \\ \cdots\ \cdots\ \cdots & \cdots & \cdots & \cdots & \cdots & \cdots \\ \cdots\ \cdots\ \cdots & \cdots & \cdots & \cdots & \cdots & \cdots \\ [Z_{am1}][Z_{am2}][Z_{am3}] & \cdots [Z_{p_{mm}}] & \cdots & & [Z_{am(n-1)}] & [Z_{ann}] \end{bmatrix} \begin{bmatrix} \begin{bmatrix} I_{qs1}{}^{a} \\ I_{ds1}{}^{a} \\ I_{qr1}{}^{a} \\ I_{dr1}{}^{a} \end{bmatrix} \\ \begin{bmatrix} I_{qs2}{}^{a} \\ I_{ds2}{}^{a} \\ I_{qr2}{}^{a} \\ I_{dr2}{}^{a} \end{bmatrix} \\ \begin{bmatrix} I_{qs3}{}^{a} \\ I_{ds3}{}^{a} \\ I_{qr3}{}^{a} \\ I_{dr3}{}^{a} \end{bmatrix} \\ \cdots \\ \cdots \\ \cdots \\ \begin{bmatrix} I_{qsn}{}^{a} \\ I_{dsn}{}^{a} \\ I_{qrn}{}^{a} \\ I_{drn}{}^{a} \end{bmatrix} \end{bmatrix} \tag{18.20}$$

where, stator voltage and current can be written as

$$\begin{bmatrix} V_{ds_m} \\ V_{qs_m} \\ V_0 \end{bmatrix} = [Park\ Matrix] \times \begin{bmatrix} V_{Rm} \\ V_{Ym} \\ V_{Bm} \end{bmatrix} \tag{18.21}$$

$$\begin{bmatrix} I_{ds_n} \\ I_{qs_n} \\ I_0 \end{bmatrix} = [Park\ Matrix] \times \begin{bmatrix} I_{Rn} \\ I_{Yn} \\ I_{Bn} \end{bmatrix} \tag{18.22}$$

Rotor voltage and current can be written as

$$\begin{bmatrix} V_{dr\,m} \\ V_{qr\,m} \\ V_0 \end{bmatrix} = [Park\ Matrix] \times \begin{bmatrix} V_{Rm} \\ V_{Ym} \\ V_{Bm} \end{bmatrix} \tag{18.23}$$

$$\begin{bmatrix} I_{dr\,n} \\ I_{qr\,n} \\ I_0 \end{bmatrix} = [Park\ Matrix] \times \begin{bmatrix} I_{Rn} \\ I_{Yn} \\ I_{Bn} \end{bmatrix} \tag{18.24}$$

In case of poly phase induction machine, rotor circuit is shorted and hence rotor voltages are zero, (18.20) can be modified as

$$\begin{bmatrix} \begin{bmatrix} V_{qs\,1} \\ V_{ds\,1} \\ 0 \\ 0 \end{bmatrix} \\ \begin{bmatrix} V_{qs\,2} \\ V_{ds\,2} \\ 0 \\ 0 \end{bmatrix} \\ \begin{bmatrix} V_{qs\,3} \\ V_{ds\,3} \\ 0 \\ 0 \end{bmatrix} \\ \cdots \\ \cdots \\ \begin{bmatrix} V_{qs\,m} \\ V_{ds\,m} \\ 0 \\ 0 \end{bmatrix} \end{bmatrix} = \begin{bmatrix} [Z_{p_{11}}]\,[Z_{a12}]\,[Z_{a13}] & \cdots & \cdots & \cdots [Z_{a1(n-1)}] & [Z_{a1n}] \\ [Z_{a21}]\,[Z_{p_{22}}]\,[Z_{a23}] & \cdots & \cdots & \cdots [Z_{a2(n-1)}] & [Z_{a2n}] \\ [Z_{a31}]\,[Z_{a32}]\,[Z_{p_{33}}] & \cdots & \cdots & \cdots [Z_{a3(n-1)}] & [Z_{a3n}] \\ \cdots\ \cdots\ \cdots & \cdots & \cdots & \cdots\ \cdots & \cdots \\ \cdots\ \cdots\ \cdots & \cdots & \cdots & \cdots\ \cdots & \cdots \\ [Z_{am1}][Z_{am2}][Z_{am3}] & \cdots [Z_{p_{mm}}] & \cdots & [Z_{am(n-1)}] & [Z_{ann}] \end{bmatrix} \begin{bmatrix} \begin{bmatrix} I_{qs\,1}{}^{a} \\ I_{ds\,1}{}^{a} \\ I_{qr\,1}{}^{a} \\ I_{dr\,1}{}^{a} \end{bmatrix} \\ \begin{bmatrix} I_{qs\,2}{}^{a} \\ I_{ds\,2}{}^{a} \\ I_{qr\,2}{}^{a} \\ I_{dr\,2}{}^{a} \end{bmatrix} \\ \begin{bmatrix} I_{qs\,3}{}^{a} \\ I_{ds\,3}{}^{a} \\ I_{qr\,3}{}^{a} \\ I_{dr\,3}{}^{a} \end{bmatrix} \\ \cdots \\ \cdots \\ \cdots \\ \begin{bmatrix} I_{qs_n}{}^{a} \\ I_{ds_n}{}^{a} \\ I_{qr_n}{}^{a} \\ I_{dr_n}{}^{a} \end{bmatrix} \end{bmatrix} \tag{18.25}$$

Or,

$$[V_{Induction\ Machine}] = [Z_{Induction\ Machine}] \times [I_{Induction\ Machine}] \tag{18.26}$$

$$[V_{Real\ Induction\ Machine}] = \begin{bmatrix} \begin{bmatrix} V_{qs1} \\ V_{ds1} \\ 0 \\ 0 \end{bmatrix} \\ \begin{bmatrix} V_{qs2} \\ V_{ds2} \\ 0 \\ 0 \end{bmatrix} \\ \begin{bmatrix} V_{qs3} \\ V_{ds3} \\ 0 \\ 0 \end{bmatrix} \\ \cdots \\ \cdots \\ \begin{bmatrix} V_{qsm} \\ V_{dsm} \\ 0 \\ 0 \end{bmatrix} \end{bmatrix} \tag{18.27}$$

$$[I_{Induction\ Machine}] = \begin{bmatrix} \begin{bmatrix} I_{qs1}{}^{a} \\ I_{ds1}{}^{a} \\ I_{qr1}{}^{a} \\ I_{dr1}{}^{a} \end{bmatrix} \\ \begin{bmatrix} I_{qs2}{}^{a} \\ I_{ds2}{}^{a} \\ I_{qr2}{}^{a} \\ I_{dr2}{}^{a} \end{bmatrix} \\ \begin{bmatrix} I_{qs3}{}^{a} \\ I_{ds3}{}^{a} \\ I_{qr3}{}^{a} \\ I_{dr3}{}^{a} \end{bmatrix} \\ \cdots \\ \cdots \\ \cdots \\ \begin{bmatrix} I_{qs_n}{}^{a} \\ I_{ds_n}{}^{a} \\ I_{qr_n}{}^{a} \\ I_{dr_n}{}^{a} \end{bmatrix} \end{bmatrix} \tag{18.28}$$

$$[Z_{Induction\ Machine}] = \begin{bmatrix} [Z_{p11}]\,[Z_{a12}]\,[Z_{a13}] & \cdots & \cdots & \cdots\,[Z_{a1(n-1)}] & [Z_{a1n}] \\ [Z_{a21}]\,[Z_{p22}]\,[Z_{a23}] & \cdots & \cdots & \cdots\,[Z_{a2(n-1)}] & [Z_{a2n}] \\ [Z_{a31}]\,[Z_{a32}]\,[Z_{p33}] & \cdots & \cdots & \cdots\,[Z_{a3(n-1)}] & [Z_{a3n}] \\ \cdots\ \cdots\ \cdots & \cdots & \cdots & \cdots\ \cdots & \cdots \\ \cdots\ \cdots\ \cdots & \cdots & \cdots & \cdots\ \cdots & \cdots \\ [Z_{am1}][Z_{am2}][Z_{am3}] & \cdots\,[Z_{pmm}] & & \cdots\,[Z_{am(n-1)}] & [Z_{ann}] \end{bmatrix} \tag{18.29}$$

a. Activity Based Model of an Ideal Poly phase Induction Motor If an ideal poly phase induction motor does not produce harmonics, then all active impedance matrices will become null. Thus the stator components will carry the harmonics which have been supplied by the source which can be represented only by passive impedance. Then (18.26) becomes

$$\begin{bmatrix} \begin{bmatrix} V_{qs1} \\ V_{ds1} \\ 0 \\ 0 \end{bmatrix} \\ \begin{bmatrix} V_{qs2} \\ V_{ds2} \\ 0 \\ 0 \end{bmatrix} \\ \begin{bmatrix} V_{qs3} \\ V_{ds3} \\ 0 \\ 0 \end{bmatrix} \\ \cdots \\ \cdots \\ \begin{bmatrix} V_{qsm} \\ V_{dsm} \\ 0 \\ 0 \end{bmatrix} \end{bmatrix} = \begin{bmatrix} [Z_{p11}] & [0] & [0] & \cdots & \cdots & [0] \\ [0] & [Z_{p22}] & [0] & \cdots & \cdots & [0] \\ [0] & [0] & [Z_{p33}] & \cdots & \cdots & [0] \\ \cdots & \cdots & \cdots & \cdots & \cdots & \cdots \\ \cdots & \cdots & \cdots & \cdots & \cdots & \cdots \\ [0] & [0] & [0] & [0] & [0] & [Z_{pnn}] \end{bmatrix} \begin{bmatrix} \begin{bmatrix} I_{qs1}{}^{a} \\ I_{ds1}{}^{a} \\ I_{qr1}{}^{a} \\ I_{dr1}{}^{a} \end{bmatrix} \\ \begin{bmatrix} I_{qs2}{}^{a} \\ I_{ds2}{}^{a} \\ I_{qr2}{}^{a} \\ I_{dr2}{}^{a} \end{bmatrix} \\ \begin{bmatrix} I_{qs3}{}^{a} \\ I_{ds3}{}^{a} \\ I_{qr3}{}^{a} \\ I_{dr3}{}^{a} \end{bmatrix} \\ \cdots \\ \cdots \\ \cdots \\ \begin{bmatrix} I_{qs_n}{}^{a} \\ I_{ds_n}{}^{a} \\ I_{qr_n}{}^{a} \\ I_{dr_n}{}^{a} \end{bmatrix} \end{bmatrix} \tag{18.30}$$

In (18.26), voltage$[V_n]$ and current$[I_n]$ have the same frequency of order n and $[Z_{pnn}]$ decides the amplitude of the voltage waveforms. Thus this equation corresponds to the case of uniform air-gap and sinusoidal distribution of flux and also rotor slip frequency is not being reflected in the stator voltage. If the supply voltage consists of only fundamental frequency then the equation will be$[V_1] = [Z_{p11}][I_1]$, which, is obviously not a real case.

b. Activity Based Model for a Real Poly Phase Induction Motor From (18.26) active model for real induction machine can be written as

$$[V_{Real}] = [Z_{Real}] \times [I_{Real}] \tag{18.31}$$

where,

$$[I_{\mathrm{Real}}] = [I_{Induction\ Machine}] \tag{18.32}$$

$$[V_{\mathrm{Real}}] = [V_{Induction\ Machine}] \tag{18.33}$$

$$[Z_{\mathrm{Real}}] = [Z_{Induction\ Machine}] \tag{18.34}$$

Like ideal machine, in a real induction machine rotor voltages are zero. Also, mesh connection of a balanced three phase system cancels all possibilities of third harmonics and hence. Thus, from (18.29) and (18.34), $[Z_{\mathrm{Real}}]$can be written as

$$[Z_{Real}] = \begin{bmatrix} [Z_{p11}] & [Z_{a12}] & [0] & [Z_{a14}] & [Z_{a15}] & [Za_{16}] & [Z_{a17}] & \cdots & \cdots & \cdots & [Z_{a1(n-1)}] & [Z_{a1n}] \\ [Z_{a21}] & [Z_{p22}] & [0] & [Z_{a24}] & [Z_{a25}] & [Za_{26}] & [Z_{a27}] & \cdots & \cdots & \cdots & [Z_{a2(n-1)}] & [Z_{a2n}] \\ [0] & [0] & [0] & [0] & [0] & [0] & [0] & \cdots & \cdots & \cdots & [0] & [0] \\ [Z_{a41}] & [Z_{a42}] & [0] & [Z_{p44}] & [Z_{a45}] & [Za_{46}] & [Z_{a47}] & \cdots & \cdots & \cdots & [Z_{a4(n-1)}] & [Z_{a4n}] \\ [Z_{a51}] & [Z_{a52}] & [0] & [Z_{a54}] & [Z_{p55}] & [Za_{56}] & [Z_{a57}] & \cdots & \cdots & \cdots & [Z_{a5(n-1)}] & [Z_{a5n}] \\ [Z_{a61}] & [Z_{a62}] & [0] & [Z_{a64}] & [Z_{a65}] & [Z_{p66}] & [Z_{a67}] & \cdots & \cdots & \cdots & [Z_{a6(n-1)}] & [Z_{a6n}] \\ [Z_{a71}] & [Z_{a72}] & [0] & [Z_{a74}] & [Z_{a75}] & [Za_{76}] & [Z_{p77}] & \cdots & \cdots & \cdots & [Z_{a7(n-1)}] & [Z_{a7n}] \\ \cdots & \cdots & \cdots & \cdots & \cdots & \cdots & \cdots & \cdots & \cdots & \cdots & \cdots & \cdots \\ \cdots & \cdots & \cdots & \cdots & \cdots & \cdots & \cdots & \cdots & \cdots & \cdots & \cdots & \cdots \\ [Z_{am1}] & [Z_{am2}] & [0] & [Z_{am4}] & [Z_{am5}] & [Za_{m6}] & [Z_{am7}] & \cdots & [Z_{pmm}] & \cdots & [Z_{am(n-1)}] & [Z_{ann}] \end{bmatrix} \tag{18.35}$$

In most of the cases even harmonics are not generated inside an induction motor and they may be present if they are supplied by the source. Thus, (18.35) can be modified as

$$[Z_{Real}] = \begin{bmatrix} [Z_{p11}] & [0] & [0] & [0] & [Z_{a15}] & [0] & [Z_{a17}] & \cdots & \cdots & \cdots & [Z_{a1(n-1)}] & [Z_{a1n}] \\ [0] & [Z_{p22}] & [0] & [0] & [0] & [0] & [0] & \cdots & \cdots & \cdots & [0] & [0] \\ [0] & [0] & [0] & [0] & [0] & [0] & [0] & \cdots & \cdots & \cdots & [0] & [0] \\ [0] & [0] & [0] & [0] & [0] & [0] & [0] & \cdots & \cdots & \cdots & [0] & [0] \\ [Z_{a51}] & [0] & [0] & [0] & [Z_{p55}] & [0] & [Z_{a57}] & \cdots & \cdots & \cdots & [Z_{a5(n-1)}] & [Z_{a5n}] \\ [0] & [0] & [0] & [0] & [0] & [0] & [0] & \cdots & \cdots & \cdots & [0] & [0] \\ [Z_{a71}] & [0] & [0] & [0] & [Z_{a75}] & [0] & [Z_{p77}] & \cdots & \cdots & \cdots & [Z_{a7(n-1)}] & [Z_{a7n}] \\ \cdots & \cdots & \cdots & \cdots & \cdots & \cdots & \cdots & \cdots & \cdots & \cdots & \cdots & \cdots \\ \cdots & \cdots & \cdots & \cdots & \cdots & \cdots & \cdots & \cdots & \cdots & \cdots & \cdots & \cdots \\ [Z_{am1}] & [0] & [0] & [0] & [Z_{am5}] & [0] & [Z_{am7}] & \cdots & [Z_{pmm}] & \cdots & [Z_{am(n-1)}] & [Z_{ann}] \end{bmatrix} \tag{18.36}$$

If the possibility of supply of even harmonics by source is neglected, $\left[Z_{p22}\right]$ becomes a null matrix and then (18.36) can be modified as

$$[Z_{\mathrm{Real}}] = \begin{bmatrix} [Z_{p11}] & [0] & [0] & [0] & [Z_{a15}] & [0] & [Z_{a17}] & \cdots & \cdots & \cdots & [Z_{a1(n-1)}] & [Z_{a1n}] \\ [0] & [0] & [0] & [0] & [0] & [0] & [0] & \cdots & \cdots & \cdots & [0] & [0] \\ [0] & [0] & [0] & [0] & [0] & [0] & [0] & \cdots & \cdots & \cdots & [0] & [0] \\ [0] & [0] & [0] & [0] & [0] & [0] & [0] & \cdots & \cdots & \cdots & [0] & [0] \\ [Z_{a51}] & [0] & [0] & [0] & [Z_{p55}] & [0] & [Z_{a57}] & \cdots & \cdots & \cdots & [Z_{a5(n-1)}] & [Z_{a5n}] \\ [0] & [0] & [0] & [0] & [0] & [0] & [0] & \cdots & \cdots & \cdots & [0] & [0] \\ [Z_{a71}] & [0] & [0] & [0] & [Z_{a75}] & [0] & [Z_{p77}] & \cdots & \cdots & \cdots & [Z_{a7(n-1)}] & [Z_{a7n}] \\ \cdots & \cdots & \cdots & \cdots & \cdots & \cdots & \cdots & \cdots & \cdots & \cdots & \cdots & \cdots \\ \cdots & \cdots & \cdots & \cdots & \cdots & \cdots & \cdots & \cdots & \cdots & \cdots & \cdots & \cdots \\ [Z_{am1}] & [0] & [0] & [0] & [Z_{am5}] & [0] & [Z_{am7}] & \cdots & [Z_{pmm}] & \cdots & [Z_{am(n-1)}] & [Z_{ann}] \end{bmatrix} \tag{18.37}$$

18.8 Discussion

At first, a passivity-based model (PBM) has been developed; its equivalent circuit and layer-based representation of the passive impedances have been drawn. Discussing its limitations, an activity-based model (ABM) has been introduced in presence harmonics. Its equivalent circuit, model and layer based representation of active impedances have been developed. On the basis of the developed models, a case study on activity-based model has been made for poly-phase induction machine. ABM has been developed for an ideal and a real induction machine.

References

[1] Carbone, R., Menniti, D., Morison, R.E., Testa, A.: Harmonics and inter-harmonic distortion modeling in multi-converter system. IEEE Trans. Power Deliv. **10**(3), 1685–1692 (1995)
[2] Chattopadhyay, S., Mitra, M., Sengupta, S.: Passive and active model for harmonics in an induction motor. Accepted in POWERCON 2008, International Conference on power system organized by IEEE in New Delhi, on 12–15 October 2008
[3] Xu, W., Ahmed, E., X. Zhang, X.: Measurement of harmonic impedances: Practical implementation issues and solutions. IEEE Trans. Power Deliv. **17**(1), 210–216 (2002)
[4] Lamedica, R., Prodenzi, A., Tironi, E., Zaninelli, D.: A model of large load areas for harmonic studies in distributions networks. IEEE Trans. Power Deliv. **12**(1), 418–425 (1997)

Index

DOI 10.1007/978-94-007-0635-4,

The manufacturer's authorised representative in the EU is Springer Nature Customer Service Centre GmbH, Europaplatz 3, 69115 Heidelberg, Germany. If you have any concerns regarding our products, please contact ProductSafety@springernature.com

Printed and bound by CPI Group (UK) Ltd, Croydon, CR0 4YY
30/09/2026
02225510-0003